D0024961

The Human and Environmental Impact of Fracking

The Human and Environmental Impact of Fracking

How Fracturing Shale for Gas Affects Us and Our World

Madelon L. Finkel, Editor

Public Health Issues and Developments

AN IMPRINT OF ABC-CLIO, LLC
Santa Barbara, California • Denver, Colorado • Oxford, England

Library of Congress Cataloging-in-Publication Data

The human and environmental impact of fracking : how fracturing shale for gas affects us and our world / Madelon L. Finkel, editor; foreword by Seth B.C. Shonkoff, PhD, MPH.

pages cm. – (Public health issues and developments)

Includes bibliographical references and index.

ISBN 978-1-4408-3259-8 (alk. paper) — ISBN 978-1-4408-3260-4 (ebook)

1. Hydraulic fracturing—Environmental aspects. 2. Hydraulic fracturing—Health aspects. I. Finkel, Madelon Lubin, 1949-

TD195.G3H86 2015

363.738'7–dc23 2014041811

ISBN: 978-1-4408-3259-8

EISBN: 978-1-4408-3260-4

19 18 17 16 15 2 3 4 5

This book is also available on the World Wide Web as an eBook.
Visit www.abc-clio.com for details.

Praeger
An Imprint of ABC-CLIO, LLC

ABC-CLIO, LLC
130 Cremona Drive, P.O. Box 1911
Santa Barbara, California 93116-1911

This book is printed on acid-free paper ∞

Manufactured in the United States of America

To my late father, Ralph Lubin,
who so enjoyed hiking in upstate New York.

Contents

Foreword

Since the early 2000s, the use of hydraulic fracturing and directional drilling as a means to extract natural gas and oil from shale source rocks and other tight geological formations has grown rapidly. Since then, more than 30 states in the United States have opened their doors to this controversial unconventional gas extraction (UGE) process. UGE has created both excitement and fear in areas across the United States (especially in areas where drilling is planned) and increasingly around the world. Proponents cite the benefits of natural gas compared with coal for electricity generation, the purported abundance of the resource, and the potential economic opportunities it affords on the local and national level. Conversely, opponents cite concerns about the climate impacts of associated methane and carbon dioxide emissions, impacts on water quantity and quality, air-quality degradation, and increased population health burdens. Questions regarding the technically recoverable quantities of natural gas from unconventional reservoirs also continue to be points of disagreement.

UGE—most notably the forms that require higher fluid volumes—is a highly industrial process, requiring land, large quantities of water, chemicals, and ancillary infrastructure to transport, process, and dispose of materials and manage the sizable waste stream. Central to the construction of sound energy policy is the tenet that rigorous scientific inquiry and moral and ethical considerations should inform such decisions. These considerations are particularly salient in cases with persistent data gaps and high economic, community, and environmental and public health stakes, as is the case for UGE. For instance, although the assertion that every gas well poses risks to drinking water supplies is unfounded, statements that gas development poses no risk to drinking water supplies are also erroneous.

Since the passage of the 2005 Energy Policy Act, the oil and gas industry has enjoyed a variety of loopholes and exemptions in most of the important federal environmental laws aimed to protect human health and the environment, including the Safe Drinking Water Act, the Clean Air Act, the Clean Water Act, and the Resource Conservation and Recovery Act. Although this relaxed federal regulatory environment reduces costs and operational restriction of UGE, it has also created obstacles to environmental monitoring and data collection. The dearth of monitoring data has slowed the ability of scientists, policymakers, and the general public to have access to sufficient information to make informed decisions about when, where, how, and if UGE should or should not be deployed.

Concomitant with the rapid increase in UGE is an increase in scientific data focusing on potential environmental, public health, and climate impacts. Of the roughly 350 peer-reviewed journal articles published directly on the subject to date, almost two-thirds were published only since the beginning of 2013, and nearly 100 of these publications since the beginning of 2014. Yes, over the past 5 years, much has been learned about the potential benefits and risks of UGE. Yet even with this surge in peer-reviewed literature, the scientific community continues to play catch-up with the rapid growth of this industry. Significant data gaps in our understanding of the potential risks of UGE for the environment and public health remain.

In this edited volume, many of the leading thinkers in the field have articulated, with the best available knowledge, science, and moral and ethical considerations, many of the issues pertinent to the discussion of UGE. They also explore how to approach critical questions about our energy future in an evolving but still quite incomplete informational context. The chapters in this edited volume will undoubtedly play a role in the scientific, policy, and community debates on the understanding of UGE. As a Western and global society, we find ourselves at a critical juncture with respect to environmental sustainability and climate stability. Many purport that gas is a key feature of a greener future, while others assert that the development and use of natural gas only exacerbates existing climate, environmental, and public health burdens. Readers of this book will be better equipped to develop an informed opinion on one of the largest environmental, social, economic, and ethical controversies of our time.

Seth B.C. Shonkoff
PSE Healthy Energy
University of California, Berkeley
Lawrence Livermore National Laboratory

Acknowledgments

This book could not have been produced without the contributions of the authors who so generously took the time to research and write their respective chapters. Each was asked to prepare a chapter because of his or her expertise on the subject and work in the area of unconventional gas development. Collectively, this book brings together the "best evidence" presented by these experts on a politically charged issue. "Fracking" is not all good, nor is it all bad. The authors understand this and present their evidence in an unbiased, objective manner, regardless of what their personal opinion of "fracking" may be.

Many thanks to my colleague and research associate, Jake Hays, for his time and effort spent preparing the book for publication. We have coauthored many articles, and I am constantly struck by his insight and knowledge of the subject. He is also a superb proofreader.

Many thanks to my editor, Debbie Carvalko, of Praeger Publishers, who invited me to be a series editor in Public Health Issues and Developments and convinced me that this book must be written. Her support and faith in my being able to deliver the goods on time was reassuring. This is the fourth book I have worked on with her, and I look forward to many more.

Introduction

Madelon L. Finkel

Looking back in time, it is clear that petroleum (oil) and gas were crucial for the development and maintenance of an industrial civilization, and this has not changed appreciably over the decades. By the early 20th century, petroleum was the most valuable commodity traded on world markets. Today, both oil and gas are vital to many industries and account for a large percentage of the world's energy consumption. Wars have been fought, fortunes made (and lost), geopolitical alliances forged, and places transformed by the newfound wealth and political power conferred by oil and gas. Both are vital to many industries and account for a large percentage of the world's energy consumption. Thus, it should come as no surprise that the search for new means of extracting oil and gas from the earth has assumed huge importance not only in the United States but around the world.

EARLY DAYS OF OIL AND GAS EXPLORATION

The history of the petroleum industry in the United States dates from the 19th century following the discovery of oil at Oil Creek, Pennsylvania (near Titusville) in 1859. George Bissell and Edwin L. Drake made the first successful use of a drilling rig on a well drilled to produce oil.[1] The Drake Well, as it was called, is considered to be the first commercial oil well in the United States, which provided the impetus for a huge wave of investment in oil drilling, refining, and marketing. Over the ensuing decades, there was a rapid expansion of oil drilling from the Appalachian Basin, the leading oil-producing region in the United States through 1904, to Oklahoma, East Texas, North Louisiana, the Gulf Coast, North Dakota,

California, and Alaska. For much of the 19th and 20th centuries the United States was the largest oil-producing country in the world, and in 2010 it regained that position after surpassing Saudi Arabia and Russia in oil production.[2]

The first well specifically intended to obtain natural gas was dug by William Hart in 1821 in Fredonia, Chautauqua County, New York.[3] Hart, referred to as the "father of natural gas" in America, noticed gas bubbles rising to the surface of a creek. He dug 27 feet into gas-bearing shale and piped the natural gas to a nearby inn, where it was burned for illumination; the gas-lit streets of Fredonia became a tourist attraction.[4] However, without an appropriate means of transporting the gas, natural gas was allowed to vent into the atmosphere, was burnt, or was simply left in the ground. Pipeline construction was a game-changer, because it made the transportation of natural gas possible.

Not many people today know who Floyd Farris or J.B. Clark are, but both men were instrumental in advancing techniques that are now used to coax oil and gas from fields that would have been bypassed as being commercially nonproductive or uneconomical to drill. Both worked for the Stanolind Oil and Gas Corporation in the 1940s, and their experiments helped open up a new way to use fluid pressure to mine petroleum and natural gas, including hydraulic fracturing, or the fracturing of rock by a pressurized liquid.[5] Farris is credited with being the first to experiment with the use of hydraulic fracturing in the Hugoton gas field in Grant County, Kansas. Gelled gasoline (essentially napalm) mixed with sand from the Arkansas River was injected 2,400 feet into the gas-producing limestone formation. Clark published a paper in which he described the process, and that same year, 1949, not only was a patent on the process issued, but an exclusive license was granted to Halliburton Oil Well Cementing Company. That year Halliburton established the first two commercial hydraulic fracturing treatments in Oklahoma (Stephens County) and Texas (Archer County).[6] In just two years (1947–1949) experimental use of hydraulic fracturing led to commercially successful applications of the technique.

The process of hydraulic fracturing involves injecting millions of gallons of water, chemical additives, and a proppant (sand and/or silica) at high pressure into the wellbore to create small fractures in the rock formations that allow natural gas (or petroleum) to be released. The fracturing fluid, consisting of water, sand, and chemical additives, creates cracks in the rock. The proppant keeps the fractures from closing. In 1968, Pan American Petroleum, in Stephens County, Oklahoma, implemented what is now known as "massive hydraulic fracturing" (e.g., high-volume hydraulic fracturing).[7] This technique involved injecting massive amounts of fluids and proppants into the wellbore, which enabled the capture of large volumes of gas-saturated sandstones that were uneconomical to drill because of low permeability. Massive hydraulic fracturing was the drilling method of choice through the 1970s.

CONVENTIONAL VERSUS UNCONVENTIONAL DRILLING TECHNIQUES

For many years conventional, or low-volume, hydraulic fracturing was used to extract oil and gas from low-permeability formations (e.g., shale gas, tight gas, tight oil, and coal seam gas). Through the 1970s, wells were drilled vertically; horizontal drilling was unusual until the 1980s.[8] Vertical drilling is fairly straightforward. With horizontal drilling, however, a wellbore is drilled vertically for thousands of feet and then directionally (i.e., horizontally) also for thousands of feet. A horizontal well is able to reach a much wider area of rock and the natural gas that is trapped within the rock; thus, a drilling company using the horizontal technique can reach more energy with fewer wells. By the 1990s, the technology advanced to the point where horizontal drilling became economical to employ. In the late 1980s, the first horizontal wells were drilled in the Texas Gulf Coast and the Austin Chalk and the Barnett Shale located in northeast Texas.[9]

In addition to vertical versus horizontal drilling, a distinction needs to be made between conventional versus unconventional drilling techniques. Conventional gas drilling applies to oil and gas that can be extracted by the natural pressure of the wells and pumping or compression operations. That is, as long as the oil and gas can flow easily from reservoirs, it is considered conventional. Reservoirs are typically porous sandstone, limestone, or dolomite rocks. Unconventional natural gas drilling is considered "unconventional" because the natural gas has not migrated from the source rock into a reservoir but remains trapped within the source rock. Unconventional reservoirs include shale rock, tight sands, and coal beds (e.g., coal-bed methane). Shale, for example, has high porosity and low permeability, making the extraction of shale gas using conventional drilling methods not feasible.

THE BASICS OF HYDRAULIC FRACTURING

Geologists were aware that there were huge reserves of natural gas in shale rock with permeability too low to recover the gas in an economical manner. In 1997, Mitchell Energy developed a hydraulic fracturing technique, slickwater fracturing, that made the extraction of shale gas economical. George P. Mitchell has been called the "father of fracking" because of his role in developing this technique.[10] Briefly, the process of hydraulic fracturing involves injecting millions of gallons of water, chemical additives, and a proppant at high pressure into the wellbore to create small fractures in the rock formations to allow natural gas (or oil) to be released. The fracturing fluid varies in composition depending on the type of fracturing used, the conditions of the well, and the concentration of chemicals used. The mixture of chemicals used by drilling companies is considered to be proprietary, and a listing of them does not have to be made public.

When the pressure is released, gas flows up the production casing, where it is collected, processed, and sent through transmission pipelines to market. Along with the gas, fluid also returns to the surface (known as flowback, produced water, or wastewater) and contains not only the chemical additives used in the drilling process but also heavy metals, radioactive materials, volatile organic compounds (VOCs), and hazardous air pollutants such as benzene, toluene, ethylbenzene, and xylene (BTEX). VOCs, a group of carbon-based chemicals, are groundwater contaminants of particular concern because of the potential of human toxicity and a tendency for some compounds to persist in and migrate with groundwater into the drinking-water supply. Each VOC has its own toxicity and potential for causing different health effects. A proportion of flowback and produced waters are treated; however, many of the chemicals remain because treatment facilities are unable to screen for and eliminate these compounds.[11] Containment and storage of these returned waters remains a huge issue, which will be discussed in subsequent chapters.

Natural gas extraction using high-volume slickwater hydraulic fracturing from clustered multiwell pads using long, directionally drilled laterals (known by its popular name "fracking"), is an unconventional natural gas extraction process that is currently the focus of controversy. Throughout this book, the process is referred to as "unconventional gas extraction" (UGE).

NATURAL GAS DEVELOPMENT: UNITED STATES AND ABROAD

The U.S. Energy Information Administration (EIA) forecasts that global energy consumption will grow 56% between 2011 and 2040. Almost 80% of that energy demand will be satisfied by fossil fuels, raising carbon emissions from 32 billion to 45 billion tons per year.[12] Much of the growth in energy consumption occurs in countries outside the Organization for Economic Cooperation and Development—that is, in emerging nations, particularly China and India. The best way to abate this trend is to find a replacement for coal.

The world's insatiable appetite for energy has helped fuel the exploration of alternative energy sources. China retains its position as the world's largest energy consumer, followed by the United States, India, Russia, and Japan.[13] Although oil and coal remain the predominant energy sources worldwide—34% and 30%, respectively—the search for cleaner, safe, efficient, economic alternatives is a high priority for both industrial and emerging nations. Natural gas (24% of the world's energy source), hydropower (6% of the world's energy source), and nuclear energy (5% of the world's energy source) are promoted as "clean" energy alternatives, albeit with pros and cons to each of these energy sources.[14] It is natural gas, however, that is receiving the most attention and interest.

The world's largest producer of natural gas is not Russia or any other gas-rich country; it is the United States. UGE is viewed as a "game-changer" for the United States, a way for the country to become independent from foreign oil. The United States is home to some major shale plays with trillions of cubic feet of shale gas.[15] The Marcellus Shale, which spans West Virginia, New York, Pennsylvania, Ohio, and Maryland, is the largest shale gas field in the United States, followed by the Barnett Shale, covering at least 24 counties in North Texas. Both are the most active shale plays in the United States. The Haynesville Shale, primarily in Louisiana and East Texas, and the Eagle Ford Shale in South Texas collectively contain trillions of cubic feet of recoverable natural gas as well. Other large natural gas shale plays in the United States include the Bakken Shale (North Dakota), Uinta Basin (western Colorado and eastern Utah), Antrim (northern Michigan), and Monterey Shale (California).[16]

Canada has an active hydraulic fracturing program in many of its provinces, although recently environmentalists and legal experts have criticized the federal government's decision to leave toxic fracking chemicals off a list of pollutants going into Canada's air, land, and water.[17] Other countries are only starting to develop shale gas reserves, and these remain largely in the exploration phase. China, for example, has 31 trillion cubic meters of recoverable shale gas, which is estimated to be the largest technically recoverable shale gas resource of any country in the world.[18] However, the two main shale gas plays are in the western part of the country, where pipelines are essentially nonexistent. Furthermore, despite its vast shale gas resources, China currently lacks the technological expertise to develop these resources because the gas is significantly deeper than that in the United States.

Poland, which imports more than 80% of its natural gas, much of it from Russia, allegedly has substantial shale gas reserves. Given its dependence on imported natural gas, the Polish government has shown strong support for unconventional gas exploration.[19] The government is offering attractive fiscal incentives for shale gas development, and Chevron has permits to explore more than 1 million acres in southeast Poland.[20] Mozambique, Tanzania, and Kenya have rich deposits, and each country is attracting interest among the world's largest energy companies.[21]

England has been moving ahead to drill for natural gas despite the two small earthquakes that rocked Lancashire County, which are attributed to natural gas operations. Public Health England issued a draft report in 2013 focusing on air and water quality and the uncertainties that surround the public health implications of extracting shale gas.[22] Although the report concluded that the health, safety, and environmental risks associated with UGE can be managed effectively in the United Kingdom, outside experts politely disagree with its conclusion.[23] Nevertheless, the British government appears to be intent on moving forward with drilling despite the furious objections among those opposed to UGE.

Since the 1960s, the Netherlands has been a large producer and consumer of natural gas thanks to the development of the Groningen gas field, the largest of its kind in Western Europe. In 2012, however, the government instituted a temporary ban on hydraulic fracturing pending results from a detailed investigation of the environmental effects of shale gas extraction. In August 2013, findings from the investigation led to the conclusion that the environmental risks associated with extraction are manageable, so long as the correct guidelines are in place.[24] At present, all drilling is on hold, and no new licenses for the extraction of shale gas will be issued until a formal decision is made by the government; therefore, no extraction of shale gas is expected in the Netherlands in the near future.

In 2011, France became the first country to ban unconventional drilling for natural gas, followed by Bulgaria. In 2013, the European Parliament voted by a narrow margin to force energy companies to carry out in-depth environmental audits before they would be allowed to use hydraulic fracturing to extract shale gas.[25] The audits would focus on the direct and indirect effects on human health, animals, land, water, and climate. South Africa's moratorium on UGE was lifted in 2012; antifracking activists are threatening a preemptive injunction against exploratory drilling if the nation's government does not place a moratorium on exploration licenses for the desert region of Karoo.[26] The U.S. EIA estimates that South Africa has a substantial reserve of technically recoverable reserves.[27]

SURGE IN UGE DEVELOPMENT

It is estimated that there are trillions of cubic meters of recoverable shale gas that could be extracted by UGE. In the United States, UGE is now the method of choice to extract natural gas as well as oil. As a proportion of the nation's overall gas production, shale gas increased from 4% in 2005 to 24% in 2012.[28] Yet UGE is a source of considerable controversy. Well-designed studies are urgently needed to provide the scientific evidence to support and/or to refute arguments put forth by both sides of the debate.[29]

PROS AND CONS OF NATURAL GAS AND UGE

Although natural gas is being touted as an energy source that is "cleaner" than other fossil fuels and that emits less carbon dioxide than coal or oil when combusted, there is the "methane issue." Carbon dioxide comes from burning fossil fuels such as coal. Methane, however, is the principal component of natural gas and is a significant greenhouse gas, which contributes to climate change/global warming. Although methane breaks down more quickly in the atmosphere compared with carbon dioxide, it is a much more potent greenhouse gas. There is an ongoing debate and discussion focusing on the potential acceleration of global warming that would occur should there be a widespread increase in UGE activities. Figuring out how to deal with methane that is emitted into the atmosphere throughout the various stages of natural gas extraction and production, as well as figuring out how to contain the leaking of methane gas into the groundwater and aquifers, are challenges that must be addressed.

The potential economic benefits of UGE are tremendous, but there is a "dirty downside" to this energy source that must be acknowledged. There are many legitimate areas of concern surrounding this process, including the potential for harm to the environment (air, water, soil, climate change), to human and animal health, and to the social and economic infrastructure of community.

Proponents of UGE make the argument that natural gas as a source of energy is attractive from an economic and environmental perspective. Drilling for natural gas will help the United States significantly reduce its dependence on foreign oil and would even put the country in the enviable position of being a major exporter of gas. UGE requires a large workforce, which would help reduce unemployment and would provide economic benefits to the areas where drilling is to take place. Natural gas releases two to three times less carbon into the atmosphere than coal, and it releases far less particulate matter as well. It also burns cleaner than coal.

Opponents make the case that the harms of UGE far outweigh the benefits. UGE will add considerably to global warming primarily because of the methane that is released from the process. UGE requires a tremendous amount of water; given the severe drought in some parts of the United States, this becomes a significant issue to address. As stated previously, the injection fluid contains toxic chemicals—including benzene, a known carcinogen, and other compounds harmful to humans and animals as well as to the environment. Scientific studies conducted to identify these chemicals have been predominantly limited to the few chemicals that were either voluntarily provided or collected from evaporation pits and blowouts of wells. The fact is that industry does not have to disclose its "proprietary chemical composition" because of exemptions granted in the Safe Drinking Water Act (e.g., the Halliburton Loophole). Consequently, there is quite a bit that remains unknown about the chemical makeup of fracking fluids.

The structural integrity of wells can and does fail. If cement does not bond properly with the walls of the well, contaminants and methane can and do leak into water supplies. Oil and gas wells routinely leak, allowing for the migration of natural gas and toxic substances as well as methane. Wells have blowouts. Spills are common.

By far, one of the most critical issues related to UGE is the management (storage, treatment, disposal) of flowback fluids. This fluid is highly toxic and, if not disposed of properly, poses irreversible harm to the environment, including polluting drinking water wells, streams, rivers, aquifers, and the like. Increases in seismic activity have been recorded and attributed to the deep-injection of flowback and produced water from hydraulically fractured wells.[30] Air pollution, too, is a serious concern. Harm to human and animal health is an issue that must be investigated in an empirical, methodologically sound manner. Economic benefits to the towns located near drilling may not necessarily materialize, and an increase in crime and substance abuse has been seen in areas being drilled. Weak regulations fail to ensure appropriate monitoring of the negative consequences of the process.

These issues are discussed in depth and in an evidence-based manner in this book. The chapters, written by experts in the field, include the most current evidence, pro and con, to help the reader understand the issues from many different perspectives.

Impact on Human and Animal Health

The paucity of scientific evidence looking at the public health impact of UGE complicates the issue. It is difficult, indeed unwise, to formulate policy and regulations in a vacuum absent data. Although there have been anecdotal reports of adverse health effects ranging from minor to serious among those living in areas where UGE is ongoing, objective and well-designed epidemiologic studies have not been conducted. Finkel and Law have called for using the precautionary principle, which asserts that the burden of proof for potentially harmful actions rests on the assurance of safety in areas of scientific uncertainty—*primum, non nocere* (i.e., first do no harm).[31] Inherent in the principle is that preventive action should be taken in the face of uncertainty, the burden of proof should be shifted to the proponents of an activity, alternatives to possibly harmful actions need to be explored, and there should be increased public participation in decision making.[32] Chapters 1, 2, and 3 present a discussion on the potential health impacts of UGE on humans and animals. Chapter 4 presents a discussion on Health Impact Assessments, which have not been routinely conducted but are very important for planning and policy decision making. Chapter 5 focuses on natural gas development and its effect on air quality and human health.

While adverse health effects may appear fairly quickly after exposure, others take more time to develop (e.g., cancers; harm to the reproductive, endocrine, and nervous systems; and delayed developmental effects).

Certainly the potential for harm will vary by proximity to drilling, length of time of exposure, the route of exposure, the safety culture of the drill operator, among other concerns. *Jerome Paulson and Veronica Tinney* (Chapter 1) present a discussion of occupational exposures related to UGE with a focus on population health exposures resulting from UGE processes. They present an overview of toxicological principles, risk assessment, potential pathways of exposure to populations from UGE, and potential health impacts associated with chemicals that are known to be used in UGE.

To better understand risk assessment, one needs to assess and quantify the impact of toxic chemicals on organisms, populations, and communities. One must determine the potential pathways in which species may be affected using routes of exposure, the organisms of concern, and anticipated end points, including hazard identification, dose–response assessment, exposure assessment, and risk characterization. The authors conclude that as the pace of drilling continues to increase, epidemiologic studies on the health impacts of UGE are urgently needed.

What effect does UGE have on the human endocrine system, and why is it important to know about this? In simple terms, the endocrine system influences almost every cell and organ in and function of the body. Hormone action plays a key role in regulating mood, growth and development, tissue function, metabolism, and sexual function and reproductive processes, as well as maintaining homeostasis and affecting the brain and behavior. *Adam Law* (Chapter 2) addresses the important topic of endocrine disrupting chemicals (EDCs).

Endocrine disruptors are chemicals that may interfere with the body's endocrine system and produce adverse developmental, reproductive, neurological, and immune effects in both humans and animals. Exposure to endocrine disruptors can occur through direct contact with pesticides and other chemicals or through ingestion of contaminated water or food, inhalation of gases and particles in the air, and through the skin. EDCs are only now being recognized as a consequence of UGE operations. Not only can EDCs cause changes in physiological systems, they can also affect growth and development. The epigenetic effects of EDCs must also be understood. Of key concern is how best to mitigate the potential adverse effects of UGE on human health. Unfortunately, as of this writing, research is hampered by an incomplete listing of chemicals used in UGE. Understanding the implications of EDCs on human health would be an important and valuable next step.

In the absence of well-designed studies on human health, animal studies can shed light on the potential harmful effects of UGE. Like a canary in the coal mine, animals and wildlife can be used as sentinels to foreshadow impacts to human health. *Michelle Bamberger and Robert Oswald* (Chapter 3) focus on the potential for harm to food animals, companion animals, and wildlife, as well as the implications for food safety. Food safety is a serious consideration, particularly because well sites are drilled in the middle of cornfields or near ponds or streams that serve as sources of water for cattle and other food animals; pipelines, processing plants, and compressor stations are often built where cattle graze and deer and other wildlife roam. In addition, the practice of land farming (a process in which drilling waste or wastewater is disposed on farmland, thus introducing toxic chemicals, including radioactive compounds, into the soil) has received only limited attention.

The authors conducted an exhaustive search of available research on domestic and companion animals. Findings indicate that a proportion of animals living near unconventional oil and gas operations suffer from a set of acute symptoms that are similar to their human counterparts. These symptoms are known collectively as "shale gas syndrome" and commonly affect the respiratory, gastrointestinal, dermatological, neurological, and vascular systems, but particularly the reproductive system. Reproductive problems may yet appear in humans, but these symptoms (abortions, stillbirth, failure to breed and cycle) are likely appearing acutely in these animal sentinels due to longer exposures and shorter gestation times. Research on wildlife has concentrated on habitat choices and indicates that animals tend to avoid living, breeding, and nesting in areas near unconventional operations, likely because of the noise and traffic associated with drilling operations. Although there are few definitive answers, many of the studies discussed in their review raise serious issues that need further investigation.

Ironically, the sage-grouse, a small bird found primarily in the western part of the United States, is at the center of a brewing controversy. The grouse is totally dependent on sagebrush-dominated habitats, which happen to be located near gas and oil fields and wind farms. Like the spotted owl, which was the center of debate over timber production on federal lands, the grouse is a factor in today's energy production debate. Loss of habitat due to energy production is threatening the grouse's existence, and environmentalists posit that the only way to save the grouse is to give it an endangered species designation, which essentially would restrict the use of land for energy production, including wind farming.[33] No decision will be made until September 2015 regarding the endangered species designation, but discussion is ongoing as to how the bird can coexist with energy development.

Health Impact Assessments

While drilling for oil and gas continues in the United States and is being seriously considered in other parts of the world, few assessments of the impact on the environment and health have been made before drilling commences. While environmental impact assessments are conducted to assess potential environmental impacts of a proposed plan or action, most do not take into account the wider determinants of health. Health Impact Assessments are defined as a combination of procedures, methods, and tools through which a policy, program, or project may be judged as to its potential effects on the health of a population, as well as the distribution of those effects within the population. *Liz Green* (Chapter 4) presents a comprehensive overview of the fundamentals of health impact assessments and how they should be performed.

Green writes that a major objective or purpose of an health impact assessment is to inform and influence decision making; however, it is not a decision-making tool per se. Health Impact Assessment is a process that considers to what extent the health and well-being of a population may be affected by a proposed action, be it a policy, program, plan, or project. It provides a systematic, objective, yet flexible and practical way of assessing potential positive and negative health impacts associated with a particular activity. It also provides an opportunity to suggest ways in which health risks can be minimized and health benefits maximized. Health Impact Assessments should be viewed as a valuable instrument to support effective decision making, particularly in regard to UGE. To date, few Health Impact Assessments have been conducted before, during, or after drilling. This situation needs to be changed to provide needed information about the potential for harm to the environment and to human and animal health.

Effect on Air Quality and Human Health

The process of extracting and transporting oil and gas is dirty, messy, and polluting to the environment if not done correctly, cleanly, and carefully. The process also involves massive amounts of chemicals that can have an adverse effect not only on air quality but also on human health. Researchers from the University of Colorado, *Nathan P. De Jong, Roxana Z. Witter, and John L. Adgate*, have prepared a scholarly piece (Chapter 5) on the impacts UGE has on air quality and health. To truly understand the impact UGE has on the environment and health, the authors make the case that one needs to understand the phases of the production process to obtain a clearer picture of the sources of chemical pollution. They present the six main categories of air pollutants (petroleum hydrocarbons, silica, diesel particulate matter, radiation, hydrogen sulfide, and ozone) as well as the potential health hazards associated with these pollutants. They also review the current, limited information about human exposure and epidemiological data related to these exposures.

The authors present a chemical primer listing the release of chemicals at each stage of production; most are known to be highly toxic and carcinogenic. Health hazards associated with air pollutants include diesel exhaust, especially PM2.5, from heavy trucks and diesel powered generators; VOC emissions, in particular BTEX, from flowback operations; waste and liquid hydrocarbon storage tanks; compressor and purification equipment and leaks; releases from transportation pipelines; criteria pollutants NOx and SOx from gas flares and diesel combustion; and nuisance dust (PM10) from traffic on unpaved roads. Emissions of ozone precursors (VOC and NOx) may lead to increased ozone formation in regions where there is heavy UGE activity. The health effects associated with these hazards range from cardiac and pulmonary effects (PM2.5, ozone), adverse birth outcomes (BTEX, ozone), short-term irritant and neurological effects (BTEX, ozone), and the psychological effects of uncertainty about UGE exposures. Density of wells; proximity to homes, schools, and other community institutions; and cumulative emissions in regional air sheds are contributing factors to community health risk. Susceptible populations include children and fetuses, the elderly, those with chronic illnesses, and outdoor workers.

Clearly there are potentially serious risks to health from the UGE process, and much more research needs to be done to fully understand the short- and long-term consequences of exposure (direct or indirect) to UGE activities. Improved exposure assessments and well-designed health studies will help frame the discussion about how to mitigate the potential for harm to human health as well as to the environment and climate.

Climate Change: A Global Perspective

Philip Staddon and Michael Depledge present a global perspective of the implications of UGE on climate change. In Chapter 6, the authors present the case, citing overwhelming evidence, that climate change is being driven primarily by human activities, including the release into the atmosphere of greenhouse gases, carbon dioxide (CO_2), and methane (CH_4), as a result of burning fossil fuels and land use changes. However, shale gas is portrayed as being "environmentally friendly" by the oil and gas industry and by governments that continue to support fossil fuel use. Proponents argue that CO_2 emissions from shale gas are lower than for coal and oil and that shale gas is a logical transition fuel to CO_2 emission-free alternative energy sources. However, there is a fallacy in this position. Shale gas actually represents an additional source of fossil fuel greenhouse gas emissions. Unconventional extraction of shale gas releases large quantities of methane into the atmosphere, and high levels of methane undermine the notion that natural gas is "good for the environment." The authors stress that methane leakage ("fugitive methane") is an important issue that must be addressed, given that fugitive methane emissions are many times more potent than carbon dioxide at trapping heat.

The authors discuss the concepts of mitigation and adaptation, the two main policy responses to climate change. Climate change mitigation refers to efforts to reduce or prevent emission of greenhouse gases. Examples of mitigation include switching to low-carbon energy sources, such as renewable and nuclear energy. Mitigation aims to control a global phenomenon, and therefore requires international cooperation for a successful outcome. Adaptation seeks to lower the risks posed by the consequences of climatic changes. Contrary to mitigation, adaptation must be delineated and implemented at the local level.

The chapter presents evidence of global warming's effect on the natural ecosystem, agriculture, health, weather patterns, and animal and bird habitats. The authors posit that based on the available data on climate change and greenhouse warming, there needs to be a concerted global effort to switch to carbon-neutral and low carbon energy alternatives to limit the "need" for fossil fuels and move to sustainable future energy sources that do not contribute to global warming.

Community Considerations

The effect of UGE on a community is an aspect that tends to be overlooked. The UGE process involves transporting the materials needed to drill to the well site, erecting the well platform, building compressor stations, and transporting the gas and flowback fluids from well sites—all of which can be hugely disruptive to the local communities located in areas where drilling is planned. Workers are brought into the community and need housing, among other basic needs. In Chapter 7, *Kathryn Brasier and Matthew Filteau* present an overview of the social and community impacts of UGE by summarizing the social science literature, with a particular focus on recent research in the Marcellus Shale region.

Brasier and Filteau discuss the "boomtown model," the main theoretical approach used to understand social impacts on extractive communities. The boomtown model has provided a critical frame of reference for researchers studying the social and economic impacts of contemporary shale-based oil and gas development focusing on critical impacts, including economic change, population change, housing impacts, institutional change (e.g., education, local government, human services, and health care services), crime, community relationships and conflict, and landscape change. It is important to acknowledge that no two communities experience energy development equally. In fact, population density, history with extractive industries, and type of extraction may affect how residents perceive energy development and social change in their communities.

An increase in population to the rural communities can be disruptive in many ways. A lack of affordable, quality housing has been identified as one of the most critical early impacts of UGE. Furthermore, an increase in population increases the demand for services provided by local governments, including human services, physical infrastructure (roads, water, sewer), and land use planning. Often, local governments need to provide these services with few additional resources. Crime is another byproduct of energy boomtowns. Crime may increase because of several factors, including an overall increase in population; an increase in young (usually single) men in the community; greater wealth in the community, creating a disparity among the local population and the workers brought in to work; and so forth.

A prominent area of research is how UGE influences residents' perceptions of their communities and the social relationships among community members. A consistent concern raised by study participants has been how the costs and benefits of development are distributed among residents in the region, with a particular concern about how development may polarize the "haves" and the "have-nots." Studies have looked at the potential economic benefits to a community and have found mixed results. Economic impact studies in the Marcellus Shale region, for example, have largely focused on job growth and have resulted in widely varying estimates of jobs associated with UGE.

History shows that "booms" are usually followed by "busts." Many rural communities that depend on natural resource extraction fall into continual boom-and-bust cycles. Studies suggest that a substantial proportion of the increased wages are going to non-local (in- and out-of-state) workers, also raising questions about the extent to which communities are benefitting from the development. The authors put forth four questions that challenge us to modify or develop a new model to understand the community impacts of energy development. This model emphasizes how the spatial, temporal, and cultural context of development shapes local impacts.

Economic Footprint of Shale Gas: Boom–Bust Cycles

To fully understand the implications of high-volume hydraulic fracturing on the U.S. economy, we need to look beyond the well pad. Specifically, what can previous experience tells us about the regional economic impacts of extraction-based resource development? What types of local economies are affected by the industrial processes connected to hydraulic fracturing? How are the differences among those places likely to alter the nature and extent of the impacts? To what extent is hydraulic fracturing affecting regional economies where no shale gas or oil development is taking place? In Chapter 8, *Susan Christopherson* discusses the economic impact of resource extraction on regional economies with particular attention to what has been learned about UGE economies in U.S. shale plays. An examination of the different types of regions affected by UGE is presented, including how differences among those regions may affect the range, visibility, and intensity of economic impact. She shows that the footprint of UGE is local, regional, and national. The drilling may be in one small rural area, but the ripple effect goes far beyond that. Although the well pad may be the locus of production, the environmental and economic costs of servicing the well site are distributed in a complex production chain that stretches across the United States.

Much has been written about the "boom-and-bust" cycle. Essentially, natural resource development—including unconventional gas extraction—is positive for some segments of the population (e.g., mineral rights owners, some businesses) and negative for others (e.g., renters, land owners without mineral rights, businesses in competing industries). When the commercially viable resources are depleted, drilling ceases, either temporarily or permanently, and there is an economic "bust" as businesses and personnel connected to resource extraction leave the community. Mineral rights owners may continue to derive royalties from their leases, but the impact of those royalties on the regional economy is unclear. Mineral rights leaseholders may not reside in the region or may invest rather than spend their royalties or may spend their royalties outside the region. So, in addition to issues related to the boom–bust nature of economic development in UGE regions, there are complex questions about how the boom period economic benefits are distributed in the population and geographically.

UGE industrial facilities create a wide range of intersecting environmental, economic, and social stressors, all of which have implications for the regional economy and its existing industries. Christopherson discusses the impact oil and gas drilling has had on communities in Wyoming, North Dakota, and in the Marcellus Shale play, including Pennsylvania, New York, and Ohio. Much of the toxic byproducts of drilling in Pennsylvania, for example, end up in the southern tier of New York State where drilling is banned, as well as in Ohio.

Legal and Regulatory Issues

In Chapter 9, attorneys *Kate Sinding, Daniel Raichel, and Jonathon Krois* present a clear and concise overview of the legal and regulatory aspects of shale gas development. Their discussion of the web of federal, state, and local laws and regulations will help the reader gain an understanding of whose laws and regulations take precedent over others. The role of the federal government in regulating shale gas development is actually relatively limited, as the states act as the primary regulators. That being said, several major laws have an impact on the regulation of shale gas development, including the Safe Drinking Water Act, the Clean Water Act, the Clean Air Act, the Resource Conservation and Recovery Act (the primary federal law designed to ensure the safe management of solid and hazardous wastes "from cradle-to-grave"—through generation, transportation, treatment, storage, and disposal—and thus to prevent the creation of new toxic waste sites), the Emergency Planning and Community Right-to-Know Act (a process for informing people of chemical hazards in their communities), and the Emergency Planning and Community Right-to-Know Act (a process for informing people of chemical hazards in their communities).

However, in the grand scheme of things, the role of the federal government in regulating shale gas development is actually relatively limited, primarily because of a network of interrelated exemptions from the nation's major federal environmental laws, including the exemption of hydraulic fracturing from regulation by the leading federal environmental regulator, the Environmental Protection Agency (EPA). Probably the most famous federal exemption is that which precludes the EPA from regulating hydraulic fracturing under the Safe Drinking Water Act.

Most significant regulation of shale gas development comes from state law, and the states have nearly unfettered authority to regulate it. With shale gas extraction currently underway in states ranging from the Northeast through the Midwest, the South, and the western United States, the result is a patchwork of state laws that vary widely in their scope, stringency, and enforcement. State regulations, for example, cover testing for gas, well spacing, setbacks, casing and cementing to protect underground water supplies, on-site storage of hydraulic fracturing and drilling fluids and fracking waste, disposal of waste, chemical disclosure requirements, testing water supplies, reporting and remediation of spills, and issues related to property rights surrounding shale gas development. Not surprisingly, rules and regulations vary greatly from state to state. In addition, states regulate to address the wide array of potential environmental impacts associated with shale development, including those relating to water, air, wildlife, habitat, and waste.

Most UGE development occurs on land within the jurisdiction of a local government. Not surprisingly, the recent nationwide surge in shale gas production has produced a parallel flood of municipal laws attempting to address local issues and in the process has raised questions as to the role of local governments. Often at stake is whether local governments are allowed any input over where and how such development occurs in their communities, or whether those decisions rest entirely with state officials. Local zoning and land use laws are greatly influential in allowing (or not allowing) gas development. Municipal laws, too, may affect shale gas operations by preventing them entirely. For example, the Court of Appeals in the State of New York (the state's highest court) recently recognized the authority of municipalities to completely exclude oil and gas drilling using zoning, declaring that current state law does not "oblige" municipalities "to permit the exploitation of any and all natural resources."

Proposals to ban or restrict UGE were on the ballot in several local elections in 2014. Although the results were mixed, it is apparent that the issue to allow or ban UGE has galvanized many communities. Denton, Texas, located in the Barnett Shale, is a small city with more than 270 natural gas wells in its jurisdiction. Election results showed that a majority voted to ban hydraulic fracturing. This was a huge win for the environmentalists and is important in that no municipality in Texas had ever banned hydraulic fracturing. Local referendums banning hydraulic fracturing passed in Colorado, California, and New York (largely symbolic in that New York has a moratorium in place pending environmental and health impact studies); however, voters in Ohio defeated a proposed ban.

The authors present a cogent overview of the interplay among federal, state, and local/municipal laws and regulations, which illustrates the complexity of the tangled legal and regulatory climate.

Ethical Considerations

Concerns about the uncertainty regarding risks and benefits of drilling and well stimulation complicate discussions on ethical concerns. Ethical values, however, play an essential role not just in the evaluation of risks and potential benefits but also in the production of scientific and technological knowledge and in science-based policy. Value judgments can and should play a legitimate role in the interpretation of the scientific evidence and in policy decisions. Values may shape research agendas, influence the interpretation of the evidence, and direct the way science is used to promote particular policies. Philosopher-ethicists *Jake Hays and Inmaculada de Melo-Martín* (Chapter 10) focus on ethical issues surrounding UGE and provide an overview of the various ways in which ethical judgments play a role in the development and implementation of policy decision making.

Despite scientific uncertainty, governments need to make decisions about how to proceed with developing shale gas. Decisions about whether to allow development and implementation of shale gas are grounded on the existent scientific evidence, but they also reflect a legislature's ethical preferences to avoid false-negative or false-positive errors. A false positive occurs when a true null hypothesis is rejected, suggesting an effect exists when one does not. For instance, for the null hypothesis—shale gas development has no detrimental effects on the health of populations—a false-positive occurs when a determination is made that detrimental effects do exist, when such is not the case. When considering policy options in situations of uncertainty, policymakers must risk either rejecting a true null hypothesis or failing to reject a false null hypothesis. In other words, they risk either failing to develop shale gas when the technology is safe or developing shale gas when doing so is unsafe. Minimizing false positives lowers the possibility of restricting a harmless activity. Minimizing false negatives limits the possibility of accepting a harmful activity.

The authors explore the role of values in scientific reasoning, consider the ways in which ethical value judgments shape preferences for particular policies regarding shale gas development, and discuss specific value considerations that should be taken into account alongside traditional risk assessments for shale gas development.

Industry Perspective

There are many stakeholders with vested interests in oil and gas development, and *Dennis J. Devlin* of ExxonMobil presents a cogent overview of industry's perspective in Chapter 11. Although the focus of this book is primarily about natural gas, from an industry perspective, it is important to recognize the significance of unconventional gas and oil, with both being produced by hydraulic fracturing. Industry refers to the process as unconventional resource development (URD). From his perspective, URD is providing substantial benefits for the United States in terms of economic growth, energy security, and pollution and greenhouse gas emission reduction. As with any industrial activity, however, there is a trade-off between benefits and risks, and the industry generally strives to identify, understand, mitigate, and manage the risks of unconventional oil and gas development.

Devlin echoes the call for empirically sound epidemiologic studies. Absent such studies, the ability to draw conclusions about the potential for harm is limited. He lays out what he believes are sound guidelines to follow and offers a perspective on the American Petroleum Industry focus on community health concerns.

Industry has much to gain from expanded URD—and much to lose if the process is not done right. Devlin offers explanations to the pressing questions about the potential risks of UGE and states that in contrast to actual benefits, risks are largely hypothetical. He states that there are few examples of documented harm and that these harms are limited in scope. Still, industry participants know that certain risks exist, and there is a diligent effort in place to recognize, mitigate, and manage them. Furthermore, industry members recognize that some people are genuinely concerned about the impact of UGD in their communities. Industry must continue to enhance its efforts to communicate effectively and transparently with all stakeholders. Over time, this will help to address concerns and build trust.

The Geology of Shale

In Chapter 12, geologist *J. David Hughes* presents a sobering discourse on the sustainability of shale gas. He takes the reader through the production potential of the major oil and shale plays in the United States. Using data from the EIA, his chapter illustrates that many of the shale and oil plays are getting "played out." The Barnett, for example, which was the first shale gas play to employ high-volume, multistage, hydraulic fracturing of horizontal wells, peaked in 2011 and is now down 18% from peak. The Haynesville play, which was the largest shale gas play in the United

States when it peaked in early 2012, is now down 46%. Other plays like the Fayetteville and Woodford, which were unknown in 2006, are on a gently declining plateau. Five legacy plays, which produced 58% of U.S. shale gas production at their collective peak in August 2012 are now down 23%. Growth in shale gas production is now primarily being supported by the Marcellus and the Eagle Ford plays, with less significant production from the Bakken, Utica, and a handful of other small plays, which will also reach their peaks over the next few years.

Even so, the same agency that publishes the statistics, the EIA, is bullish on the future growth of shale gas and tight oil production in the United States. The EIA is similarly bullish on future production of oil from shale. Hughes explains that projecting future production in shale plays is a matter of determining essential play parameters such as field decline rate, well quality by area, and number of available drilling locations. Then, by estimating drilling rates, a future production profile can be developed. He presents a case study of the oldest shale gas play—the Barnett—to illustrate the salient fundamentals that ultimately control long-term production from all shale plays and concludes that emerging plays will be required to meet huge growth rates to satisfy the EIA's forecasts. Many more wells will have to be drilled in known plays over the next two decades concomitant with intense drilling in emerging plays. U.S. energy policy would be well advised to factor the realities of shale into its long-term energy planning. Shale certainly has been a short-term game-changer, but long-term realities must be considered in a sustainable energy plan.

CONCLUDING THOUGHTS

Anthony R. Ingraffea (Chapter 13) has prepared some final thoughts on the subject. Whether one agrees with him (or any of the other authors in this volume for that matter) or not, the material presented is based on current, peer-reviewed articles, special reports, and databases. The particular perspective that each author may hold is immaterial to this book, which is not a vehicle for an individual to vent his or her personal feelings about unconventional gas development.

All of the authors invited to prepare a chapter are experts in their field. After reading each chapter, one can see that common themes are interwoven throughout. All of the authors acknowledge that there are potential benefits and potential risks to UGE. All have called for more empirical research to better assess the potential for harm to the environment and to human and animal health. The need to develop alternatives to coal and oil is acknowledged globally; yet to blindly view unconventional natural gas development as "our savior" is short-sighted and could end up being potentially more harmful to the climate, the environment, and human health.

There are some serious issues that must be addressed before a push to drill. Consideration must take into account the geological potential (estimated oil and gas reserves), the estimates of capital expenditures necessary to extract the oil and gas, and the environmental and human costs of UGE. The economics and politics of unconventional gas extraction further complicate matters. Cool heads must prevail if we in the United States—and globally—are to "get this right." The primary purpose of this book is to educate, inform, and raise issues for discussion. Enjoy the read.

NOTES

1. Owen EW. *Trek of the Oil Finders*. Tulsa, OK: American Association of Petroleum Geologists, 1975.

2. Smith G. U.S. seen as biggest oil producer after overtaking Saudi Arabia (July 4, 2014). http://www.bloomberg.com/news/2014-07-04/u-s-seen-as-biggest-oil-producer-after-overtaking-saudi.html (accessed July 4, 2014).

3. NaturalGas.org. *History*. naturalgas.org/overview/history (accessed July 29, 2014).

4. Owen, *Trek of the Oil Finders*, op. cit.

5. Montgomery CT, Smith MB. Hydraulic fracturing. History of an enduring technology. *Journal of Petroleum Technology* (2010): 26–40.

6. Ibid.

7. Law BE, Spencer CW. Gas in tight reservoirs—an emerging major source of energy. In *The Future of Energy Gasses* (U.S. Geological Survey Professional Paper 1570), ed. DG Howell, 1993. pubs.er.usgs.gov/publication/pp1570 (accessed July 27, 2014).

8. Bell CE. *Effective Diverting in Horizontal Wells in the Austin Chalk*. Paper presented at the Society of Petroleum Engineers, 1993. http://www.onepetro.org/mslib/servlet/onepetropreview?id=00026582 (accessed July 23, 2014).

9. Ibid.

10. Zuckerman G. *How fracking billionaires built their empires*. November 5, 2014. http://qz.com/144435/how-fracking-billionaires-built-their-empires (accessed July 27, 2014).

11. Shonkoff SB, Hays J, Finkel ML. Environmental public health dimensions of shale and tight gas development. *Environmental Health Perspectives*. Published electronically April 2014. Dx.doi.org/10.1289/ehp.1307866.

12. U.S. EIA. *Annual Energy Outlook 2012*. http://www.eia.gov/forecasts/archieve.aeo12 (accessed April 22, 2012).

13. U.S. EIA. *Lower 48 States Shale Plays*. http://www.eia.gov/oil_gas/rpd/shale_gas.pdf (accessed August 23, 2014).

14. U.S. EIA. *International Energy Outlook 2013*. http://www.eia.gov/forecasts/ieo/more_highlights.cfm (accessed July 23, 2014).

15. U.S. EIA. U.S. to become world's leading producer of oil and gas hydrocarbons in 2013. http://www.eia.gov/todayinenergy/detail.cfm?id=13251# (accessed July 27, 2014).

16. FracTracker Alliance. Over 1.1 million active oil and gas wells in the US (May 4, 2014). http://www.fractracker.org/2014/03/1-million-wells (accessed July 27, 2014).

17. The Canadian Press. Feds criticized for leaving fracking chemicals off pollutant list. July 23, 2014. http://www.theepochtimes.com/n3/812939-feds-criticized-for-leaving-fracking-chemicals-off-pollutant-list (accessed July 23, 2014).

18. Varun Chandra A. China's shale gas push could be huge for U.S. companies. The Motley Fool, July 14, 2014. http://www.fool.com/investing/general/2014/07/24/chinas-shale-gas-push-could-be-huge-for-us-compani.aspx (accessed July 23, 2014).

19. Shale development in Poland. Vinson & Elkins Shale & Fracking Tracker. http://fracking.velaw.com/shale-development-in-Poland (accessed July 29, 2014).

20. Chevron teams up with PGNiG to explore shale gas in southern Poland. *Business Review + Poland Weekly Newsletter* (No. 015). December 16, 2013. http://www.poland-today.pl (accessed July 27, 2014).

21. Scott M. Out of Africa (and elsewhere): more fossil fuels. *New York Times*, April 10, 2012. http://www.nytimes.com/2012/04/11/business/energy-environment/quest-for-new-fossil-fuels-goes-to-africa-and-beyond.html?pagewanted=all&_r=0 (accessed July 23, 2014).

22. Kibbe A, Cabianca T, Daraktchieva ZZ, et al. Review of the potential public health impacts of exposures to chemical and radioactive pollutants as a result of shale gas extraction. Public Health England (2013). http://www.hpa.org.uk/webc/HPAwebFile/HPAweb_C/1317140158707 (accessed July 27, 2014).

23. Law A, Hays J, Shonkoff S, Finkel ML. Public Health England's draft report on shale gas extraction. Mistaking best practices for actual practices. *BMJ* 348 (2014):g2728. doi:10.1136/bmj.g2728.

24. Bird & Bird. Shale gas in the Netherlands: key industry updates III. September 8, 2014. http://www.twobirds.com/en/news/articles/2014/netherlands/shale-gas-in-the-netherlands-key-industry-updates-iii (accessed November 27, 2014).

25. Kanter J. Europe votes to tighten rules on drilling method. *New York Times*, October 9, 2013. http://www.nytimes.com/2013/10/10/business/energy-environment/european-lawmakers-tighten-rules-on-fracking.html?_r=0 (accessed July 27, 2014).

26. Anti-fracking groups threaten South Africa exploration. Reuters, July 22, 2014. http://www.reuters.com/article/2014/07/22/safrica-fracking-idUSL6N0PX4MA20140722 (accessed July 23, 2014).

27. U.S. EIA. South Africa. http://www.eia.gov/countries/cab.cfm?fips=SF (accessed July 23, 2014).

28. Shale of the century. *The Economist*, June 2, 2012. http://www.economist.com/node/21556242 (accessed July 27, 2014).

29. Finkel ML, Hays J. The implications of unconventional drilling for natural gas: a global public health concern. *Public Health* 127, no. 10 (2013): 889–893. http://dx.doi.org/10.1016/j.puhe.2013.07.005.

30. Kim WY. Induced seismicity associated with fluid injection into a deep well in Youngstown, Ohio. *J Geophys Res* 118 (2013): 3506–3518.

31. Finkel ML, Law A. The rush to drill for natural gas: a public health cautionary tale. *American Journal of Public Health* 101, no. 5: 784–785. doi: 10.2105/AJPH.2010.200089.

32. Goldstein BD. The precautionary principle also applies to public health actions. *American Journal of Public Health* 91, no. 9 (2001): 1358–1361.

33. Cardwell D, Krauss C. Frack quietly, please: sage grouse is nesting. *New York Times*, July 19, 2014. http://www.nytimes.com/2014/07/20/business/energy-environment/disparate-interests-unite-to-protect-greater-sage-grouse.html (accessed July 19, 2014).

Chapter 1

Potential and Known Health Impacts Associated with Unconventional Natural Gas Extraction

Jerome A. Paulson and Veronica Tinney

INTRODUCTION

Natural gas extraction using high-volume, slickwater hydraulic fracturing from long laterals from clustered multiwell pads, known by its popular name "fracking," is an unconventional natural gas extraction process that is currently the focus of controversy. Throughout this chapter, the process will be referred to as unconventional gas extraction (UGE). We avoid the use of the term "fracking" because it is one step in this multistep process, and our concerns about potential health impacts extend beyond just hydraulic fracturing and include other aspects of the process.

Because UGE is such a widespread practice in some areas of the United States and in other countries, primary care providers and other health professionals need to be aware of the process, how it is carried out, the chemicals used and produced, potential routes of human exposure, and potential human health outcomes. Health care providers need to be in a position to respond to questions from individuals who may have been exposed to the toxic chemicals used in UGE and to be able to respond effectively to community concerns about the potential health impacts of UGE should the process be in place or proposed for the area where the provider lives or practices.

Unfortunately, there are few well-designed studies on the health impacts of unconventional gas extraction on the populations that work and live near UGE facilities. Instead, public health professionals must take what is known about the potential pathways of exposure to a population and the toxic substances involved in UGE processes and use that information to derive conclusions about potential health threats. When discussing environmental exposures, it is important to understand the route of exposure as well as the duration of exposure. The presence of a toxic substance in close proximity to a human being does not mean that an adverse health effect will occur. A toxic substance must move from its point of origin (a source of contamination), through the air or water (transported by some mechanism through an environmental media) to where it will contact a living organism (human, animal, plant), have a route into the organism (in the case of humans and animals, eating, drinking, breathing, or absorption through the skin), be distributed to the organs of the body, and then have an adverse effect on the organism. As others in this volume discuss, there are a number of toxic substances that are used in and produced by UGE processes. Some of these substances are human-made (also known as toxicants), and others are naturally occurring (known as toxins). This chapter discusses some occupational exposures related to UGE but focuses mostly on population health exposures resulting from UGE processes.

REVIEW OF TOXICOLOGICAL PRINCIPLES: THE DOSE REALLY DOES NOT MAKE THE POISON

To understand how the environmental chemicals involved in UGE affect the human population, it is essential to have a basic grasp of toxicological principles. Toxicology is the study of adverse physiochemical effects of a chemical, physical, or biological agent on living organisms and ecosystems and includes the prevention and amelioration of such adverse effects.[1] A poison is any agent capable of producing a deleterious response in biological systems. Virtually every known chemical has the potential of being poison.

Paracelsus (1493–1541), considered the father of toxicology, is reported to have said that all substances are poisons; there is nothing that is not a poison. The right dose differentiates a poison from a remedy. Although this is true in some instances, it is not always a true statement, and it is important to understand when it is or is not true. Lets use water as an example. A human being can have too little water, too much water, or just enough water. Too little or too much water certainly is toxic. Similarly, with calcium, the human body may have too little and malfunction or too much and malfunction. However, if the human body has the right amount of calcium, the body will function properly. In contrast, lead is a toxin that has no known threshold of effect. In other words, given the technology we have today, it is impossible to identify a level of lead that is low enough that adverse outcomes are avoided.

Substances generally reach the body either via the air or the water, and there may also be direct contact between the substance and the body. There must be a pathway of exposure in that the substance must have a way to reach and get into the body. Once the substance has encountered the body, it then must be absorbed for it to have any effect. The lungs, for example, present a tremendous surface area through which the substances that reach the alveoli can be absorbed. Materials can be absorbed by the gastrointestinal (GI) tract through the stomach, the small intestines, or the colon. Minor amounts of exogenous materials may be absorbed through mucous membranes, such as those of the eyes or the nose or the mouth.

Several different "doses" need to be considered when determining the actual dose of a chemical to a person. The "exposure dose" is the amount of the substance that comes in contact with the body. That substance is then absorbed through either the skin, the GI tract, or through the lungs, or less frequently, the mucous membranes of the eyes and nose, constituting the "internal dose." The substance is then distributed through the blood. Metabolism or modification of the substance, which may make it more or less toxic, can take place in the blood and often takes place in the liver. After metabolism, the substance is delivered to all of the organs of the body, called the target organ dose. Target organs may be the fat, bone, muscle, brain, or even the blood itself. For toxic substances that arrive in the body via the oral route, they undergo first-pass metabolism in the liver. However, toxic substances entering via the skin or the lungs bypass the liver, the primary site of metabolism for toxic substances, and are transported directly to target organs. Most toxic substances are excreted via the kidneys and into the urine or via the liver into the gallbladder and then out of the body in the stool. There are some volatile compounds that can be exhaled through the lungs.

Figure 1.1 illustrates the exposure pathway wherein an environmental hazard reaches a population.

Figure 1.1
Exposure pathway through which an environmental hazard reaches a population.

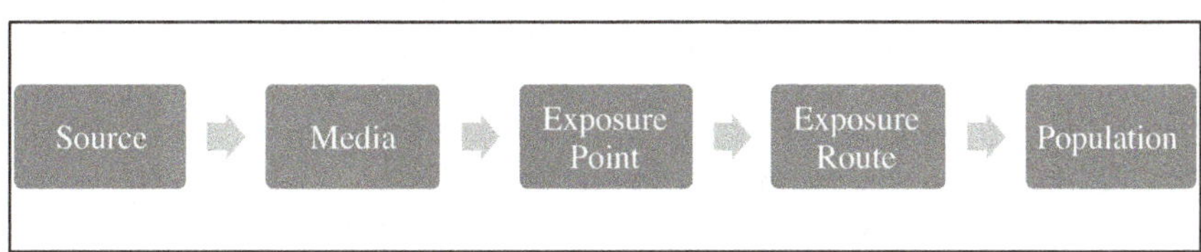

Source: Agency for Toxic Substances and Disease Registry.[2]

The dose of a particular toxic substance can be considered in three ways. These three doses may not necessarily be the same, and it is likely that not all of an exposure dose will be absorbed and become the internal dose. The first is the *exposure dose,* which is how much of a substance is encountered in the environment. The second is the *internal or absorbed dose,* which is the amount of a substance that actually gets into the body through the skin, GI tract, lungs, or mucous membranes. Absorption depends primarily on the solubility of a chemical and the different characteristics of each route of entry. Finally, there is the target organ dose, which is the amount of the toxic substance that actually reaches the organ where the insult will occur. The internal dose is often metabolized, or biotransformed, by the liver or blood before reaching the target organ. The target organ dose is often less than the exposure dose because of the liver's detoxification mechanisms, and in some instances the resulting metabolite is more toxic than the original chemical. The half-life of a compound or substance is also important to take into account when considering the toxicology of a substance. The term half-life refers to the amount of time it takes for half of the dose of the substance to exit the body. There can be different half-lives for bone, blood, the central nervous system, and other compartments in the body.

One of the best examples of these kinds of differences between the exposure dose and absorbed dose is lead. Lead also illustrates how absorption can change with the age of the subject. Children between birth and 2 years of age are estimated to absorb between 42% and 53% of the amount of lead that gets into the GI tract, whereas adults absorb approximately 7% to 15% of the same oral dose.[3] Figure 1.2 illustrates the process through which an environmental agent is distributed and absorbed by the organ or tissue it affects.

Dose–Response

Population dose–response is the change in likelihood of response with changing dose. At each dose level, there are members of the population who respond or do not respond to the dose. Often there is a dose below which there is no response, called the "no observed adverse effect level" (NOAEL). There may also be a dose, which is usually higher, in which all individuals exposed to that dose will experience an effect. The midpoint is the dose–response 50%, wherein half of the exposed individuals will experience the endpoint of interest. In some cases, the dose–response relationship may be a linear one; however, sometimes the dose–response curve is U-shaped, an inverted U-shape, or other nonmonotonic relationship. Several endocrine-disrupting chemicals have been shown to have nonmonotonic dose–response curves. Figure 1.3 illustrates the dose–response curve.

Figure 1.2
Process through which an environmental agent is distributed and absorbed by the organ or tissue it affects.

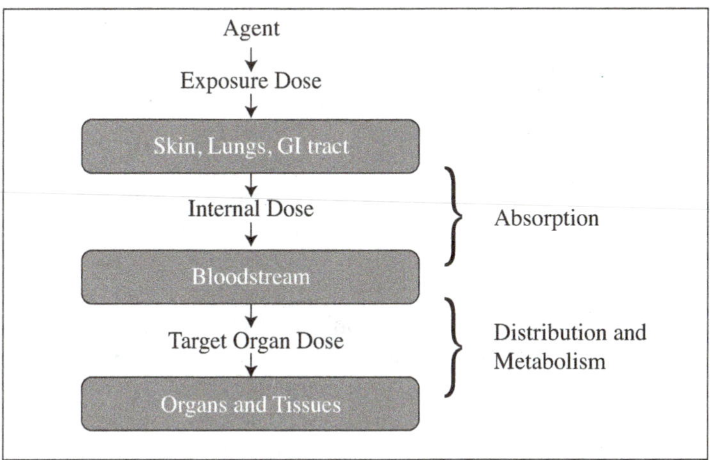

Source: Agency for Toxic Substances and Disease Registry.[4]

Figure 1.3
The dose–response curve.

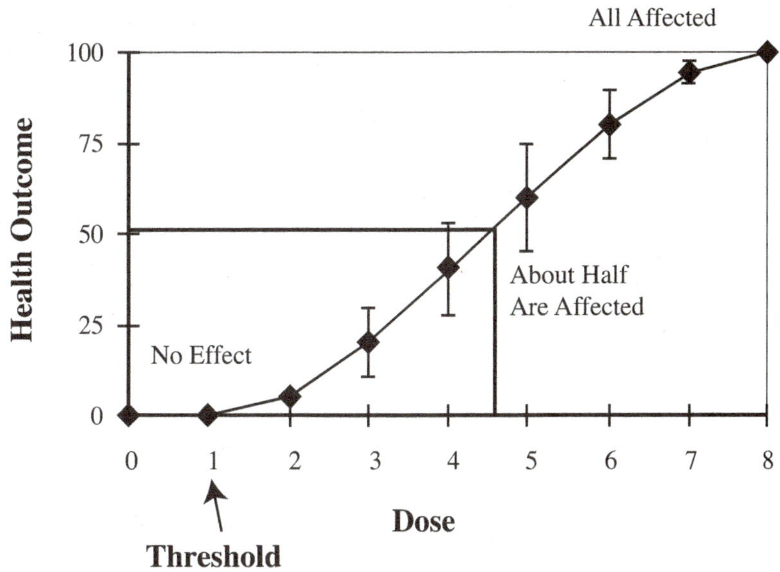

Source: Vandenberg et al.[5]

Factors Influencing Outcome Other Than Dose

The frequency of exposure influences the outcome of a dose. Consider this example: an individual drinks four beers in an hour. Alternatively, an individual drinks four beers spaced out over 4 days. The total volume of exposure is the same; however, the two scenarios will have two different outcomes. Duration of exposure also influences outcome.

Exposures are categorized as acute, subacute, subchronic, and chronic. An acute exposure occurs over a short period of time. A subchronic exposure occurs over a longer period of time, and a chronic exposure occurs continuously over a much longer period of time. The same chemical can have highly varied outcomes depending on whether the exposure is acute or chronic, with some endpoints not apparent until after repeated, long-term exposure.

Time frames defining acute, subchronic, and chronic are not specifically described, and one must look at the particular document one is reviewing and determine how the author of that document is using those terms. In animal studies, for example, acute studies are typically a one-time exposure less than 24 hours. All repeated exposures are referred to as subacute, subchronic, and chronic. Subacute is usually less than 1 month, subchronic 1 to 3 months, and chronic a year or longer.[6]

Risk Assessment

Toxicity tests are categorized depending on the outcome of interest, including reproductive and developmental toxicity, neurotoxicity, mutagenicity, oncogenicity, and immunotoxicity. To make the toxicological information meaningful, it must be put into the context of what is the overall risk. Risk is a function of hazard and exposure, and risk assessment is the qualitative and quantitative estimate of potential effects on human health from various chemical exposures. Risk assessment uses toxicological information to inform decision making. Using exposure data and the toxic effect of chemicals of interest, one determines risk. Then one must consider other issues such as sublethal effects and population or community impacts. That is, in undertaking risk assessment, one needs to assess and quantify the impact of toxic chemicals on organisms, populations, and communities. One must determine the potential pathways in which species may be affected using routes of exposure, the organisms of concern, and anticipated end points.

There are four steps in risk assessment: hazard identification, dose–response assessment, exposure assessment, and risk characterization. Figure 1.4 shows the steps in risk assessment.

Epidemiological, in vitro, and animal studies are methods used to determine whether a chemical is a hazard during the hazard identification process. "Hazard" refers to the toxic properties of the substance. Hazard identification determines whether an exposure to a hazard can induce

Figure 1.4
Steps in risk assessment.

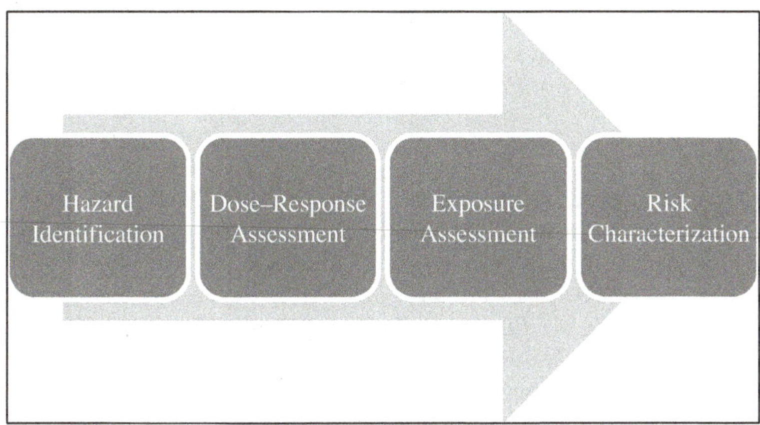

Source: James, Warren, Halmes, et al.[7]

adverse effects in a population. Dose–response assessment refers to determining the relationship between the dose of a chemical and the health outcome of interest and considers frequency and duration of exposure. Exposure assessment ascertains the routes of exposure, the duration, frequency, media, and populations that may have high exposures, such as workers. Exposure assessment may occur through direct measurement of a chemical in air, water, and soil or may be a modeled estimate of the cumulative exposure through multiple exposure mediums.

Risk characterization takes the information from the hazard identification, dose–response, and exposure assessment processes and estimates the risk in either a descriptive or quantitative manner. The characterization of risk can help inform policy decisions through risk management and risk communication. For example, the U.S. Environmental Protection Agency (EPA) often characterizes risk as a probability and has established several acceptable levels of risk for chemicals in water and air. Risk management is the process through which policy actions are chosen to control hazards identified. Risk communication is the process of making information about risk assessment available and comprehensible to stakeholders and the public. The following is a brief overview of the potential for harm from UGE. Other chapters in this book expand on these topics in greater detail.

TOXICANTS IN UNCONVENTIONAL GAS EXTRACTION

UGE involves multiple steps: pad construction, drill setup, drilling, hydraulic fracturing (or fracking), gas extraction, gas processing, well decommissioning, and land restoration. Although each of the steps in the process of UGE may have health consequences, public health concerns arise from the drilling, hydraulic fracturing, natural gas extraction, and gas processing stages. Pad construction, drill setup, well decommissioning, and land restoration have potential adverse health impacts related to workers but generally not to the public.

Air Pollution

Air pollution occurs during every stage of UGE. During site preparation heavy equipment is used to clear and prepare the well-pad site and to create new roads. Generators are set up, and there are emissions from vehicles and generators if they are diesel, as well as increased course particulate matter and dust from the new roads and increased truck traffic on the roads. During the hydraulic fracturing process, air pollutants are caused by the enormous amounts of sand and/or silica needed to fracture the well.

In an analysis of all chemicals used in UGE processes, Colborn et al. found that 37% were volatile and therefore able to aerosolize. Of these volatile chemicals, 81% were found to have adverse effects on the brain and central nervous system.[8] Aerosolized chemicals have the ability to be inhaled and absorbed directly into the bloodstream, bypassing the body's detoxifying mechanisms of the liver. Analysis of materials from a well blowout before the fracking process of the well revealed 22 chemicals used in the drilling process, of which 100% of the chemicals were found to have adverse respiratory effects.[9] Well completion was cited as the highest period of air pollution, with air pollutants such as methane emissions, hydrogen sulfide, and volatile organic compounds (VOCs) coming back to the surface in flowback water.

Well production requires the use of compressors and pumps to bring produced gas up to the surface and pressurized gas for pipeline distribution. During onsite processing, oil and water is removed from the natural gas and stored in condensate tanks. Off-site processing may require that the gas be further pressurized for longer transport. All of these processes release VOCs and methane gas and can contribute to the formation of ozone.

Air Pollution from Diesel Engines and Truck Traffic

Literally thousands of truck trips are necessary to establish a pad, set up the drilling equipment, bring workers and supplies to the pad, drill and frack the well, and remove waste from the pad and transport methane (if it is not transported in pipelines). Materials needed to construct the well, such as pipe segments, also need to be brought to the pad and can be 31 or 48 feet in length. If a well is 7,000 feet deep and travels 1 mile laterally, that requires almost 260 segments of 48 feet each. Further, trucks are used to bring large quantities of water required from the source to the well pad. In addition, thousands of tons of sand need to be trucked in.

One analysis commissioned by the New York State Department of Environmental Conservation estimated that approximately 1,975 heavy-load round-trip truck trips and 1,420 light-load round-trip truck trips are required for each horizontal well with high-volume hydraulic fracturing.[10] In Pennsylvania, it is estimated that each well requires approximately 1,235 one-way truck trips to transport water and sand, which equates to 2,262 truck trips per well and 13,572 to 18,096 trips for a multiwell pad with six to eight wells.[11] Diesel exhaust is not benign. A number of federal agencies and international bodies now classify it as "carcinogenic to humans"[12] or "reasonably anticipated to be a human carcinogen"[13] and "likely to be carcinogenic to humans."[14]

Dust and particulate matters are also created by the large volume of truck traffic. For those living along haul routes, increased truck traffic increases diesel exhaust, creates noise and vibration, and creates safety risks. In addition to truck traffic, traffic also increases from an increased population of workers commuting to and from the pads. A health impact assessment in Battlement Mesa, Colorado, estimated traffic would increase by 40 to 280 truck trips per day per pad with as many as 120 to a 150 additional workers commuting to the well pads.[15]

Air Pollution from Extraction

Methane and VOC release can occur at any stage of exploration, for example, during production through venting, flashing, and flaring or during storage and transportation through fugitive emissions.[16] Numerous pieces of industrial equipment are needed, including diesel trucks, drilling rigs, power generators, phase separators, dehydrators, storage tanks, compressors, and pipelines. Each can be a source of methane, VOCs, nitrogen oxides, particulate matter, and other gases. Methane that comes up from the well is not pure methane but is super methane-containing VOCs.

In the first few days of production, methane is not captured, and it is either burned off in the flaring process or released into the atmosphere. During this period, when the additional chemicals such as benzene, toluene, ethyl benzene, and xylene (BTEX) and hydrogen sulfide are not removed from the methane, these chemicals are also released into the atmosphere with the methane. The majority of emissions during extraction come from the flowback period, followed by pneumatic controllers, equipment leaks, and chemical pumps.[17] Once methane is captured to be stored in tanks, cylinders, pumps, and pipelines, all of those things leak in limited degrees and contribute to air pollution. Even during the use phase of methane, natural gas continues to leak from pipes and storage containers. Recently, in Boston, 3,356 methane leaks were identified exceeding 2.5 ppm[18] (a level of 2.5 ppm being the level higher than normal background levels of methane) and 5,893 above 2.5 ppm in Washington, DC.[19]

Air Pollution from Sand

Sand, usually consisting of silicon dioxide, is used as a proppant to hold open the fractures in the shale to allow the gas to escape. From a medical standpoint, this is similar to a stent used to keep a coronary artery open despite significant atherosclerosis. Silica sand is the most reportedly used material in UGE.[20] When sand is mined, transported, or used at the drill site before wetting, it can become aerosolized, presenting a hazard to the miners and workers on the pad. Aerosolized sand is a known cause of silicosis.[21] It is unknown whether there is sufficient exposure to sand used at the drill site before wetting for a long enough period of time to pose a threat to individuals off the pad site. The other hypothetical exposure to silica to individuals off the pad site is the aerosolization of silica and other particulate material by containment pond aerators, which are used to decrease the volume of flowback and produced water needed to be stored on or near the drill site. Levels of silica measured at hydraulic fracturing operations have been shown to far exceed occupational exposure limits for respirable silica. Wells can have up to 40 stages (zones) that are fractured, and the higher the stage, the more sand and water is needed.[22]

Air Pollution from Benzene

Benzene, a known carcinogen, is an organic compound found most often in air as a result of emissions from burning coal and oil, motor vehicle exhaust, and other sources. It can volatize to vapors in the air. Benzene is commonly used in the UGE process and is released from the ground, along with natural gas. At the time of writing, there are few methodologically sound studies examining the relationship between benzene exposure from UGE and adverse health outcomes. That said, there are data on perinatal exposure to benzene from exposure to petroleum refineries in Texas and child health outcomes. Two studies examined populations in residential proximity to petroleum refineries and birth outcomes in the Texas birth defect registry. The studies found that women exposed to benzene during pregnancy are more likely to have children with neural tube defects and the two most common types of leukemia.[23,24] A study in France assessed perinatal exposure to benzene by having women wear monitors to collect data on personal exposure to benzene. Women who had the most exposure to automobile and truck traffic near their homes were more likely to have children with smaller growth parameters than the women who were less exposed to traffic near their homes.[25]

A human health risk assessment of air emissions conducted by McKenzie and colleagues quantified the risk of noncancer and cancer endpoints.[26] Exposure was separated into residents less than half a mile from well pads and greater than a half mile. Exposure was then determined with ambient air samples around well pads and categorized as during the well completion phase, when at least one well was undergoing uncontrolled flowback emissions, and not during the completion phase. The results of the risk assessment found that the high exposure of the completion phase created the greatest risk due to higher exposure levels to several hydrocarbons. Residents living less than a half mile from a well had an elevated risk of both noncancer and cancer endpoints. The elevated risk for cancer was found to be 6 in 1 million for residents greater than half a mile, and 10 in 1 million for greater than half a mile, both of which are above the Environmental Protection Agency target of acceptable risk of 1 in a million. The authors found that benzene was a major component of the elevated cancer risk.

Air Pollution from Ozone

Processes of UGE create an environment in which there are multiple precursors to ozone formation. Ozone is formed when oxides of nitrogen, which can come from diesel exhaust, and VOCs, which can come from either the diesel exhaust or from the natural gas released from underground, interact with sunlight. Ground-level ozone is an irritant to the lungs.[27] Some of the health effects associated with ozone are shortness of breath, coughing, and aggravation of chronic lung diseases such as asthma, emphysema, and chronic obstructive pulmonary disease. Damage to the lungs continues even when symptoms have dissipated.

Exposure to ozone during childhood not only exacerbates asthma but can also lead to new-onset asthma and permanently affect lung function.[28] Analysis of data from the Children's Health Study looked at child lung function of 3,677 children aged 10 to 18 to determine the role of air pollution in lung development.[29] Children living within 500 meters of a freeway were found to have lesser lung function compared with their counterparts living 1500 or more meters from a freeway. Everyone loses some lung function as they age, but children with lesser lung function may be more likely to develop chronic lung diseases as adults.

Water Pollution

There are multiple ways in which water may become contaminated from UGE activities. During the injection process, well-casing failure may occur or water can migrate through fractures. On the well pad, water quality issues arise from spills or leaks from the storage of chemicals and fluid in storage tanks or containment ponds. Drilling mud, produced or flow-water,

Figure 1.5
Areas within the UGE process wherein water contamination may occur.

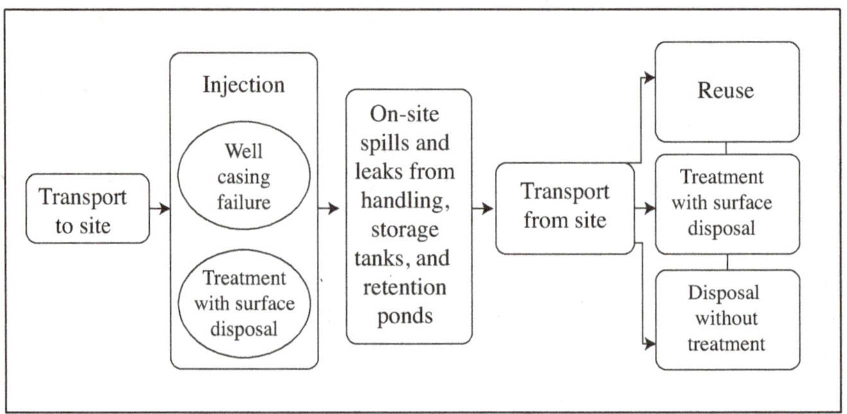

Source: Rozell and Reaven.[30]

is reused in the injection or drilling process. Figure 1.5 outlines the steps within the UGE process wherein water contamination may occur.

Methane in the Water Supply

In both Pennsylvania and Texas, there have been clear, well-documented episodes of migration of natural gas intro the drinking water supply. Researchers from Duke University sampled Pennsylvania wells and were able to demonstrate that proximity to a well increases the risk of methane in well water.[31] Isotopes of carbon in the methane in these wells showed that the methane was made thousands of years ago and therefore more likely to come from shale gas, as opposed to newer methane, which can occur with decomposing organic material.[32] The causes of gas migration and stray gas are well known, but differing geology and drilling practices make risk of this migration difficult to predict.

In addition to explosion risk, methane also causes water quality concerns because arsenic and iron can more easily dissolve in water when oxygen depletion occurs, which can result from methane oxidation by bacteria. Anaerobic bacteria can also reduce sulfate present in water to sulfide.[33] Further discussion of methane as a contributor to climate change is presented in Chapter 6.

Hydraulic Fracturing Fluid as a Contaminant

Hydraulic fracturing fluid (HFF) is a mixture of substances that is pumped underground at very high pressure to break up the gas-containing shale. In most instances, the primary component of the HFF is water mixed with a proppant and a number of chemicals. Some of the HFF, in addition to chemicals normally occurring underground that become dissolved in or mixed with the HFF, then return to the surface (e.g., flowback water). When water, sand, and chemicals are injected into the drill hole, some of the water stays underground; however, up to an estimated 90% of water used in fracking returns to the surface during the course of the extraction process. In addition to all of the materials that were injected into the water, there are additional materials that return with the fracturing fluid, including radioactive material, salts, salts of manganese, chlorides, sodium bromides, and heavy metals such as lead and arsenic.

Flowback water refers to the water that returns during the fracking process and "produced" water refers to water that returns once the well is producing gas and continues to return throughout the life of the well. Materials that would otherwise be stored in the rock and do not have a pathway of exposure to humans are released and come back with the fracturing fluids. For example, as much as 200 tons of salt can return in the flow-back water in the Marcellus.[34]

Naturally occurring radioactive material (NORM) is radiation underground in geographic formations that become technologically enhanced naturally occurring radioactive material (TENORM) when it is disturbed and has the potential for human exposure.[35] Radionuclides shown to be present in natural gas wastes include radon; ^{226}radium and ^{228}radium; and radionuclides of potassium, strontium, lead, thallium, bismuth, and thorium. Radium in flowback and produced water often incorporates into solids formed during wastewater treatment, thereby producing low-level radioactive waste.

It is the components of the HFF as well as the added material from underground that have the potential to contaminate ground or surface waters and thereby expose humans. The greatest risk of exposure to HFF fluid is when HFF flowback and produced water is stored in containment ponds, which can overflow and contaminate surface water. The composition of HFF can vary widely depending on a number of factors. HFF fluids have a variety of purposes including surfactants, acids, gelling agents, biocides, proppants, bactericides, corrosion inhibitors, stabilizers, friction reducers, among others (Table 1.1). Colburn and colleagues identified 632 chemicals as being used in the drilling, fracking, processing, and transport processes; however, this number only represents the number of chemicals in products that had listed Chemical Abstract Service numbers.[36] The same study found that for 407 of the 944 products identified (43%), there was less than 1% of the product composition available.

Data collected by the Minority Staff of the Committee on Energy and Commerce of the US House of Representatives in 2011, based on data submitted by the 14 leading oil and gas service companies, revealed the use of

Table 1.1
Hydraulic Fracturing Fluid Additives, Their Use, and Main Compound

Additive Type	Use	Main Compound
Acid	Cleans out the wellbore, dissolves minerals, and initiates cracks in rock	Hydrochloric acid
Biocide	Controls bacteria growth	Glutaraldehyde, 2,2-dibromo-3-nitrilopropionamide (DBNPA)
Breaker	Delays breakdown of the gelling agent	Ammonium persulfate
Corrosion inhibitor	Prevents corrosion of pipes	N,n-dimethyl formamide
Crosslinker	Maintains fluid viscosity as temperature increases	Borate salts
Friction reducers	Decreases pumping friction	Polyacrylamide, petroleum distillate
Gelling agents	Improves proppant placement	Guar gum, hydroxyethyl cellulose
KCl	Creates a brine carrier fluid	Potassium chloride
Oxygen scavenger	Prevents corrosion of well tubulars	Ammonium bisulfite
pH adjusting agent	Adjusts the pH of fluid to maintain the effectiveness of other components	Sodium carbonate, carbonate
Scale inhibitor	Prevents scale deposits in the pipe	Ethylene glycol
Surfactant	Winterizing agent	Isopropanol, ethanol, 2-butoxyethanol

Source: U.S. Government Accountability Office. *Oil and Gas: Information on Shale Resources, Development, and Environmental and Public Health Risks* (GAO-12-732). Washington, DC: Government Accountability Office, 2012.

more than 2,500 hydraulic fracturing products containing 750 chemicals and other components.[37] From the limited information available, it is evident that many of the substances used in hydraulic fracturing fluid are toxic, including some that are known carcinogens.[38] That being said, companies are not required to disclose the contents of their products, and listing "proprietary ingredients" is an acceptable description. As a result, the full range of chemicals that can be incorporated into HFF has never been publically revealed, making it nearly impossible to determine the extent of exposure of workers and individuals in nearby communities. This situation is untenable for first responders, emergency and primary care providers, poison control centers and pediatric environmental health specialty units, occupational medicine providers, and all other health care providers. Appropriate emergency planning is impossible, as are accurate diagnostic and therapeutic decisions about individuals who have been exposed.

Wastewater Containment, Disposal, and Treatment

Data from the Pennsylvania Department of Environmental Protection regarding disposal of wastewater from July 2012 to July 2013 found that unconventional well operators reported approximately 1.76 billion gallons of that wastewater. Of that wastewater, 68.7% was reused for purposes other than road spreading, 18% was sent to a centralized treatment plant for recycling, and 12.4% was disposed via Class II injection well disposal.[39] Although the use of pits is becoming less common, in some instances, waste from the well pad is stored in pits, also called containment ponds, before being trucked offsite. Pits can contain both flowback and produced water, drilling mud, brine, HFFs, and cuttings (i.e., metal, rock, and other shavings produced by the drill bit). If the pits are not properly lined, some of this water can seep out of the pit or overflow in heavy rains. A risk analysis of water pollution in the Marcellus shale found that the greatest uncertainty for water pollution is from wastewater disposal and for retention pond breaches resulting in large discharges from well-pad sites.[40]

Reports from pits in New Mexico identified 40 chemicals and metals in evaporation pits with 98% of the chemicals found to be listed under the U.S. EPA's Comprehensive Environmental Response, Compensation, and Liability Act (Superfund) list and 73% under the Emergency Planning and Community Right-to-Know Act reportable toxic chemicals.[41]

EPIDEMIOLOGICAL STUDIES

This chapter has focused on the pathways in which populations can be exposed to UGE processes as well as the known and potential environmental hazards associated with these processes. Environmental epidemiologic studies can help inform hazard identification in the absence of randomized control trials. Unfortunately, few epidemiologic studies have been conducted to determine the impacts of exposure to UGE processes on population health. In medicine, the gold standard for determining association is a randomized control trial (RCT); however, in public health, RCTs are often impossible and unethical.

One well-designed study focused on the relationship between maternal proximity to well operations and infant health. McKenzie and colleagues also looked at the relationship between proximity and density of gas wells to maternal address and birth defects, preterm birth, and fetal growth. In this study, two approximately even exposure groups were formed for births in rural Colorado between 1996 and 2009: zero wells within 10 miles and one or more wells within 10 miles. For women residing with one or more wells within 10 miles, women were then categorized into three groups of increasing number of wells within 10 miles. Women in the highest exposure group, with greater than 125 wells per mile, had an elevated risk of births with coronary heart disease (odds ratio 1.3, confidence interval: 1.2–1.5) and congenital defects (odds ratio 2.0, confidence interval: 1.0–3.9).[42] A monotonic relationship between increased risk for both conditions and increasing number of wells was seen. The authors cited chemicals such as benzene, solvents, and air pollutants as previously established associations between maternal exposure and coronary heart disease and congenital defects.

Mental health problems have been associated with proximity to drilling sites. Well pad operations, when set up, are industrial facilities often running 24 hours a day near homes, schools, and public areas, creating unhealthy noise levels for the surrounding area. Although noise is a part of our daily life, with typical conversations occurring at sound levels between 55 and 60 decibels (dbA), annoyance to noise can begin to occur at sound levels around 55 dbA, school performance begins to decline at 70 dbA, and sleep is disturbed at anywhere from 35 to 60 dbA. For well pads, noise levels have been shown to be 89 to 90 dbA at 50 feet from the pad, 60 to 68 dbA at 500 feet, and 63 to 54 dbA at 1,000 feet from the pad.[43] In Colorado, wells are required only to be 350 feet from areas considered high density and 150 feet from all other structures.[44] A setback of 150 to 350 feet has no health basis and is still well within the range of disturbed sleep, classroom performance, and annoyance. Further, the law only applies to new wells, not to existing wells, causing many homes, schools, and recreational areas in Colorado to be situated very close to well-pad operations.

A community study by Ferrar and colleagues found that the predominant stressor of citizens affected by shale gas drilling in Pennsylvania was a concern for their health.[45] The majority of persons interviewed felt that their health concerns were largely ignored; the most common health complaint of community members was stress. Stressors may also include odors, such as from the rotten egg smell of hydrogen sulfide released by unconventional gas extraction operations.

Community-based participatory research (CBPR) is a way to engage citizens in voicing their concerns about a subject. CBPR could be used to determine potential health impacts of an environmental hazard. Community members would be trained to act as citizen scientists and collect air and water samples, for example. A CBPR study of residents in Pennsylvania examined self-reported symptoms of residents both in close proximity to well pads, defined as less than 1,500 feet, and residency farther away, at greater than 1,500 feet.[46] The study found that residents closer to the wells reported more health symptoms, with the most common being increased fatigue, nasal and throat irritation, sinus problems, shortness of breath, headaches, and sleep disturbance. The study was also able to measure one-time water and air monitoring samples among a subset of participants and found that reported symptoms were similar to the health effects of chemicals found in the air and water monitoring tests.

POPULATION HEALTH IMPACTS

UGE may also cause more subtle changes in community health. For example, a health impact assessment done in Battlement Mesa, Colorado, found that unconventional gas extraction activities create community-wide impacts, including an increased transient worker population and a decreased use of public outdoor areas.[47] The assessment also found increased crime rates and rates of sexually transmitted infections. Other identified health impacts include increased traffic accidents, decreased use of outdoor space and reduced physical activity, increased stress, a decline of social cohesion, and strain on community resources such as health care and housing, due to an influx of workers. Chapter 7 in this volume by Brasier and Filteau expounds on this issue in greater detail.

CONCLUSIONS

This chapter has provided an overview of toxicological principles, risk assessment, potential pathways of exposure to populations from UGE, and potential health impacts associated with chemicals that are known to be used in UGE activities. Studies that examine the association between exposure and health impacts are limited, however, despite the numerous anecdotal and documented health outcomes among workers and citizens in areas of UGE and the potential for human exposure. As the pace of drilling continues to increase, systemically collected and peer-reviewed studies on the health impacts of UGE are urgently needed to confirm exposure pathways and health outcomes identified here and by others. Funding of these health studies should not be the burden of state or government agencies, universities, or citizens, but rather the companies that undertake UGE activities. The industry should be required to monitor the health of citizens in the areas where they operate and cease operations should negative health outcomes be discovered. Until such time, the health impacts of exposure to UGE will continue to play out among the populations living and working near UGE processes.

ACKNOWLEDGMENT

The U.S. Environmental Protection Agency (EPA) supports the Pediatric Environmental Health Specialty Units (PEHSUs) by providing funds to the Agency for Toxic Substance and Disease Registry (ATSDR) under Inter-Agency Agreement No. DW-75-92301301-0. Neither the EPA nor ATSDR endorse the purchase of any commercial products or services mentioned in PEHSU publications. For more information on environmental health, PEHSUs can provide consultation on UGE and pediatric environmental health issues. There are 10 PEHSUs in the United States, one for every federal region. To contact a PEHSU, find the center for your region at www.pehsu.net or by calling 1-888-347-2632.

NOTES

1. U.S. National Library of Medicine, National Institutes of Health. *ToxLearn: A Multi-Module Toxicology Tutorial*. http://toxlearn.nlm.nih.gov (accessed June 30, 2014).
2. Agency for Toxic Substances and Disease Registry. "Exposure Evaluation: Evaluating Exposure Pathways." Chapter 6 of *Public Health Assessment Guidance Manual* (2005 Update). http://www.atsdr.cdc.gov/hac/PHAManual/ch6.html (accessed June 30, 2014).
3. U.S. Environmental Protection Agency. *Review of the National Ambient Air Quality Standards for Lead: Exposure Analysis Methodology and Validation* (EPA-450/2-89-022). Washington, DC: Office of Air Quality Planning and Standards, U.S. Environmental Protection Agency, 1989.
4. Public Health Assessment Guidance Manual (2005 Update).
5. Vandenberg LN, Colborn T, Hayes TB, et al. Hormones and endocrine-disrupting chemicals: low-dose effects and nonmonotonic dose responses. *Endocrine Reviews* 33, no. 3 (2012): 378–455.
6. Eaton DL, Gilbert SG. "Principles of Toxicology." In C. D. Klaassen, ed. *Casarett and Doull's Toxicology: The Basic Science of Poisons*, pp. 11–43. New York: McGraw-Hill, 2008.
7. James R, Warren D, Halmes N, et al. "Risk Assessment." In P. Williams, R. James, S. Roberts, eds. *Principles of Toxicology: Environmental and Industrial Applications*, pp. 437–477. New York: John Wiley & Sons, 2000.
8. Colborn T, Kwiatkowski C, Schultz K, et al. Natural gas operations from a public health perspective. *Human and Ecological Risk Assessment* 17, no. 5 (2011): 1039–1056.
9. Ibid.
10. ALL Consulting. NY DEC SGEIS Information Requests: New York Department of Environmental Conservation, 2010.
11. Sibrizzi C. *Impacts of Hydraulic Fracturing on Climate Change: An Assessment of Life-Cycle Greenhouse Gas Emissions Associated with the Use of Water, Sand and Chemicals in Shale Gas Production of the Pennsylvania Marcellus Shale*. Master's thesis, The George Washington University, Washington, DC, 2014.

12. The International Agency for Research on Cancer. Diesel Exhaust Carcinogenic. www.iarc.fr/en/media-centre/pr/2012/pdfs/pr213_E.pdf (accessed July 7, 2014).

13. U.S. National Toxicology Program. www.ntp.niehs.nih.gov/ntp/roc/content/profiles/dieselexhaustparticulates.pdf (accessed July 7, 2014).

14. U.S. Environmental Protection Agency. Integrated Risk Information System (IRIS). http://www.epa.gov/iris/subst/0642.htm.

15. Witter RZ, McKenzie L, Stinson KE, et al. The use of health impact assessment for a community undergoing natural gas development. *American Journal of Public Health* 103, no. 6 (2013): 1002–1010.

16. Gilman JB, Lerner BM, Kuster WC, et al. Source signature of volatile organic compounds from oil and natural gas operations in northeastern Colorado. *Environmental Science & Technology* 47, no. 3 (2013): 1297–1305. doi:10.1021/es304119a.

17. Allen DT, Torres VM, Thomas J, et al. Measurements of methane emissions at natural gas production sites in the United States. *Proceedings of the National Academy of Sciences of the United States of America* 110, no. 44 (2013): 17768–17773. doi:10.1073/pnas.1304880110.

18. Phillips NG, Ackley R, Crosson E, et al. Natural gas leaks in Boston. *Mineralogical Magazine* 76, no. 6 (2012): 2229.

19. Jackson RB, Down A, Phillips NG, et al. Natural gas pipeline leaks across Washington, DC. *Environmental Science & Technology* 48, no. 3 (2014): 2051–2058. doi:10.1021/es404474x.

20. Colborn et al., Natural gas operations, op. cit.

21. Liu YW, Steenland K, Rong Y, et al. Exposure-response analysis and risk assessment for lung cancer in relationship to silica exposure: a 44-year cohort study of 34,018 workers. *American Journal of Epidemiology* 178, no. 9 (2013): 1424–1433.

22. Esswein EJ, Breitenstein M, Snawder J, et al. Occupational exposures to respirable crystalline silica during hydraulic fracturing. *Journal of Occupational and Environmental Hygiene* 10, no. 7 (2013): 347–356. doi:10.1080/15459624.2013.788352.

23. Lupo PJ, Symanski E, Waller DK, et al. Maternal exposure to ambient levels of benzene and neural tube defects among offspring: Texas, 1999–2004. *Environmental Health Perspectives* 119, no. 3 (2011): 397–402. doi:10.1289/ehp.1002212.

24. Whitworth KW, Symanski E, Coker AL. Childhood lymphohematopoietic cancer incidence and hazardous air pollutants in southeast Texas, 1995–2004. *Environmental Health Perspectives* 116, no. 11 (2008): 1576–1580. doi:10.1289/ehp.11593.

25. Slama R, Thiebaugeorges O, Goua V, et al. Maternal personal exposure to airborne benzene and intrauterine growth. *Environmental Health Perspectives* 117, no. 8 (2009): 1313–1321. doi:10.1289/ehp.0800465.

26. McKenzie LM, Witter RZ, Newman LS, et al. Human health risk assessment of air emissions from development of unconventional natural gas resources. *The Science of the Total Environment* 424 (2012): 79–87. doi:10.1016/j.scitotenv.2012.02.018.

27. U.S. Environmental Protection Agency. *Ground-level ozone: health effects.* http://www.epa.gov/groundlevelozone/health.html (accessed June 16, 2014).

28. Searing D, Rabinovitch N. Environmental pollution and lung effects in children. *Current Opinion in Pediatrics* 23, no. 3 (2011): 314–318. doi:10.1097/MOP.0b013e3283461926.

29. Gauderman WJ, Vora H, McConnell R, et al. Effect of exposure to traffic on lung development from 10 to 18 years of age: a cohort study. *Lancet* 369, no. 9561 (2007): 571–577. 2007. doi:S0140-6736(07)60037-3.

30. Rozell DJ, Reaven SJ. Water pollution risk associated with natural gas extraction from the marcellus shale. *Risk Analysis* 32, no. 8 (2012): 1382–1393. doi:10.1111/j.1539-6924.2011.01757.x.

31. Osborn SG, Vengosh A, Warner NR, et al. Methane contamination of drinking water accompanying gas-well drilling and hydraulic fracturing. *Proceedings of the National Academy of Sciences of the United States of America* 108, no. 20 (2011): 8172–8176. doi:10.1073/pnas.1100682108.

32. Jackson RB, Vengosh A, Darrah TH, et al. Increased stray gas abundance in a subset of drinking water wells near Marcellus shale gas extraction. *Proceedings of the National Academy of Sciences of the United States of America* 110, no. 28 (2013): 11250–11255. doi:10.1073/pnas.1221635110.

33. Vidic RD, Brantley SL, Vandenbossche JM, et al. Impact of shale gas development on regional water quality. *Science* 340, no. 6134 (2013): 1235009. doi:10.1126/science.1235009.

34. Ibid.

35. Rich AL, Crosby EC. Analysis of reserve pit sludge from unconventional natural gas hydraulic fracturing and drilling operations for the presence of technologically enhanced naturally occurring radioactive material (TENORM). *New Solutions* 23, no. 1 (2013): 117–135. doi:10.2190/NS.23.1.h.

36. Colborn et al., Natural gas operations, op. cit.

37. Minority Staff, Committee on Energy & Commerce, US House of Representatives. Chemicals Used in Hydraulic Fracturing. 2011. http://democrats.energycommerce.house.gov/sites/default/files/documents/Hydraulic-Fracturing-Chemicals-2011-4-18.pdf (accessed June 29, 2014).

38. U.S. Government Accountability Office. *Oil and Gas: Information on Shale Resources, Development, and Environmental and Public Health Risks* (GAO-12-732). Washington, DC: Government Accountability Office, 2012.

39. Sibrizzi, *Impacts of Hydraulic Fracturing on Climate Change*, op. cit.

40. Rozell and Reaven, Water pollution risk, op. cit.

41. Colborn et al., Natural gas operations, op. cit.

42. McKenzie LM, Guo R, Witter RZ, et al. Birth outcomes and maternal residential proximity to natural gas development in rural Colorado. *Environmental Health Perspectives* 122, no. 4 (2014): 412–417. doi:10.1289/ehp.1306722.

43. Oil and Gas Noise. www.earthworksaction.org/issues/detail/oil_and_gas_noise (accessed November 27, 2014).

44. Statement of Basis, Specific Statutory Authority, and Purpose New Rules and Amendments to Current Rules of the Colorado Oil and Gas Conservation Commission, 2 CCR 404-1 Cause no. 1R Docket no. 1211-RM-04 Setbacks. Colorado Oil and Gas Conservation Commission. 2012.

45. Ferrar KJ, Kriesky J, Christen CL, et al. Assessment and longitudinal analysis of health impacts and stressors perceived to result from unconventional shale gas development in the Marcellus shale region. *International Journal of Occupational and Environmental Health* 19, no. 2 (2013): 104–112. doi:10.1179/2049396713Y.0000000024.

46. Steinzor N, Subra W, Sumi, L. Investigating links between shale gas development and health impacts through a community survey project in Pennsylvania. *New Solutions* 23, no. 1 (2013): 55–83. doi:10.2190/NS.23.1.e.

47. Bureau of Land Management. *Environmental Assessment: Cache Creek Master Development Plan for Oil and Gas Exploration and Development, Garfield County, Colorado* (DOI-BLM-CO-N040-2009-0088-EA). Silt, CO: Department of Interior, Bureau of Land Management, 2009.

Chapter 2

The Public Health Risk of Endocrine Disrupting Chemicals

Adam Law

> All scientific work is incomplete—whether it be observational or experimental. All scientific work is liable to be upset or modified by advancing knowledge. That does not confer upon us a freedom to ignore the knowledge we already have, or to postpone the action that it appears to demand at a given time.
>
> —Sir Austin Bradford Hill[1]

INTRODUCTION

Has the end of cheap conventional energy arrived? Have we reached peak oil production? These are intriguing questions that must be answered in the context of geopolitical and economic realities. In today's world, growth economics continues to depend on the abundance of cheap energy. To meet the world's insatiable need for energy, other sources must be found. To fill this gap, unconventional fossil fuel extraction taps energy reserves that were previously too expensive or technically impossible to exploit. Proponents of unconventional technologies of high-volume horizontal hydraulic fracturing (HVHF) view this novel form of energy production as a solution to the world's energy problems. HVHF (hereafter referred to as unconventional gas extraction—UGE) has its strengths, but there are also some serious concerns associated with this technology.

This chapter focuses on one specific aspect of UGE: its effect on the endocrine systems of living organisms, especially humans. The integrity of multicellular organisms is dependent on a complex signaling system to integrate both short-term actions (the nervous system) and longer term processes (the endocrine system). These signaling systems evolved in a changing environment, allowing some resilience for the organisms to adapt to a variety of chemical, physical, and biological stressors. However, there are limits to this adaptive response, and the science of toxicology and the emerging field of endocrine disrupting chemicals (EDCs) represent two such limits. It should become clear to the reader that the topic has serious implications for both health and policy development.

UNCONVENTIONAL GAS EXTRACTION AND EXPOSURE PATHWAYS

In unconventional tight gas and oil reserves, the fossil fuel remains trapped in the original compacted sediment because it is unable to diffuse through more porous geology to collect in a conventional reserve. Petrochemical engineers have developed methods of releasing gas or oil droplets from these deep formations by the introduction of millions of gallons of fluids comprising proprietary chemicals mixed with sand to hydraulically fracture the formations and prop open these new channels. A substantial amount of the fluid injected comes back up as flowback wastewater that is now mixed with chemicals, hydrocarbons, and micro-organisms found in the shale.

The primary environmental impacts associated with UGE result from the use of toxic chemicals and the subsequent release of additional toxic chemicals and radioactive materials during well production. The fluid pumped out of the well and separated from oil and gas contains not only the chemical additives used in the drilling process but also heavy metals, radioactive materials, volatile organic compounds (VOCs), and hazardous air pollutants (HAPs), such as benzene, toluene, ethylbenzene, and xylene (BTEX). VOCs, a group of carbon-based chemicals, are groundwater contaminants of particular concern because of the potential of human toxicity and a tendency for some compounds to persist in and migrate with groundwater to the drinking-water supply.[2] Each VOC has its own toxicity and potential for causing different health effects.

There are numerous pathways throughout the drilling, extraction, and transporting process of UGE for the release of VOCs into the air and water. VOCs may mix with the methane and comprise a significant proportion of the produced natural gas or unintentional gas leaks (fugitive emissions). Some soluble VOCs may dissolve in the liquid phase and evaporate at a later period from evaporation pits. Therefore, a critical early step in the public health assessment process is evaluating exposure pathways.

An exposure pathway is the link between environmental releases and local populations that might come into contact with, or be exposed to, environmental contaminants.[3] A main purpose of exposure pathway evaluations is to identify likely site-specific exposure situations and answer the following questions: To what extent are individuals (and animals, for that matter) at a given site exposed to environmental contamination? Under what conditions does this exposure occur? Figure 2.1 illustrates an exposure pathway adapted from the Agency for Toxic Substances and Disease Registry.

Figure 2.1
An exposure pathway.

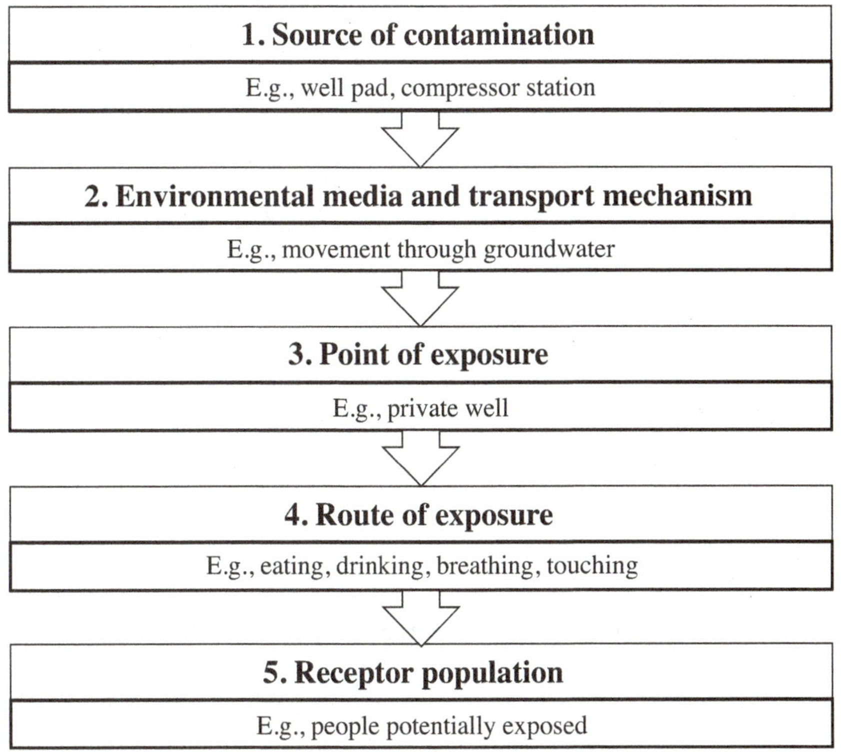

Source: Agency for Toxic Substances and Disease Registry.[4]

OVERVIEW OF THE ENDOCRINE SYSTEM

What effect does UGE have on the human endocrine system, and why is it important to know about this? In simple terms, the endocrine system influences almost every cell, organ, and function of the body. Hormone action plays key roles in regulating mood, growth, and development; tissue function; metabolism; sexual function and reproductive processes; and maintenance of homeostasis; it also has effects on the brain and behavior. Basically, the endocrine system is a collection of glands that produce hormones that regulate the body's growth, metabolism, and sexual development and function. The hormones are released into the bloodstream and transported to tissues and organs throughout the body.

Briefly, the *adrenal glands* secrete hormones that influence the body's metabolism, blood chemicals, and body characteristics, as well as the part of the nervous system that is involved in the response and defense against stress. The *hypothalamus* activates and controls the part of the nervous system in charge of involuntary body functions, the hormonal system, and many body functions, such as regulating sleep and stimulating appetite. *Ovaries and testicles* secrete hormones that influence female and male characteristics, respectively. The *pancreas* secretes a hormone (insulin) that controls the use of glucose by the body. *Parathyroid glands* secrete a hormone that maintains the calcium level in the blood. The *pineal gland* produces the serotonin derivative melatonin, a hormone that affects the modulation of sleep patterns in the circadian rhythms and seasonal functions. The *pituitary gland* produces a number of hormones that influence various other endocrine glands, and the *thymus gland* plays a role in the body's immune system. Finally, the *thyroid gland* produces hormones that stimulate body heat production, bone growth, cardiac function, and the body's metabolism.

The growth of endocrine glands, the synthesis of hormones, and their release are all tightly regulated and integrated by other hormones (i.e., pituitary hormones), the nervous system (i.e., the sympathetic and parasympathetic nervous system), metabolic molecules (i.e., free fatty acids and insulin secretion), electrolytes (i.e., calcium and parathyroid hormone regulation), and other signaling molecules (i.e., immunological or developmental signals). Diseases of the endocrine system are common, including conditions such as diabetes mellitus, thyroid disease, and obesity.

ENDOCRINE DISRUPTING CHEMICALS: WHAT IS THE SOURCE?

Any system in the body controlled by hormones can be derailed by hormone disruptors. Endocrine disruptors are chemicals that may interfere with the body's endocrine system and produce adverse developmental, reproductive, neurological, and immune effects in both humans and animals.

Exposure to endocrine disruptors can occur through direct contact with pesticides and other chemicals or through ingestion of contaminated water or food, inhalation of gases and particles in the air, and through the skin. Endocrine disruption is not considered a toxicological end point per se, but a functional change that may lead to adverse effects. The Endocrine Society recently published a new definition of Endocrine Disrupting Chemicals (EDCs) that has the virtues of being brief and clear: An EDC is an exogenous chemical, or mixture of chemicals, that interferes with any aspect of hormone action.[5]

Research on EDCs has focused on the mechanisms of hormone action of high-potency EDCs and their direct interaction with hormone receptors as well as their subsequent effect on hormone action. It is this interference with hormone receptors, resulting in either inhibition or stimulation of hormone action, that provides EDCs with effects on many different tissues, interfering with the range of processes regulated by specific hormones during the life cycle of living organisms. Of special concern are effects on early development of both humans and wildlife because these effects are often irreversible and may not become evident until later in life.[6]

Results from animal models, human clinical observations, and epidemiological studies are extremely useful, if not imperative, to assess the effect of EDCs as a significant concern to human health. Of particular concern, and pertinent to this chapter, is that chemicals used in unconventional shale gas operations have the potential to negatively affect the endocrine system and candidate EDCs. There is growing and clear evidence that EDCs have effects on male and female reproduction, breast development and cancer, prostate cancer, neuroendocrinology, the thyroid gland, metabolism and obesity, and cardiovascular endocrinology.[7] Studies in cells and laboratory animals have shown that EDCs can cause adverse biological effects in animals, and low-level exposures may also cause similar effects in human beings.[8]

The problem, however, is that for most associations reported between exposure to EDCs and physiologic or biological outcomes, the mechanism(s) of action are not well understood.[9] Furthermore, these substances are diverse and may not appear to share any structural similarity other than usually being small molecular mass compounds (<1,000 Daltons). Thus, it is difficult to predict whether a compound may or may not exert endocrine-disrupting actions.[10] This situation makes it difficult to distinguish between direct and indirect effects as well as primary and secondary effects of exposure to EDCs. Often exposure data are lacking. Fortunately, there is a growing body of evidence that has managed to accumulate credible, scientifically sound data to better understand the relationship between UGE and EDCs and human health outcomes.

Colborn and colleagues were the first to compile a database of chemicals used in UGE.[11] They abstracted information from the Medical Safety

Data Sheets, which contain information on potential hazards (health, fire, reactivity, and environmental) and how to work safely with chemical products. These sheets are prepared by the supplier or manufacturer of the material. Unfortunately, the product names do not necessarily provide detailed information of all the unique chemicals used. The lack of full disclosure by the industry of the chemicals used in drilling muds, hydraulic fracturing fluid, and other processes involved in UGE is a major obstacle to researchers. Not only has it led to a guessing game as to the chemicals used in the drilling and extraction process, it also creates an unnecessary tentativeness to conclusions reached in any study.

Colburn's group listed the chemicals by their unique Chemical Abstracts Services registry numbers (CAS) and then searched literature databases such as the Hazardous Substances Database and TOXNET, among others, to determine whether adverse biological effects were previously described for each of these chemicals. The researchers evaluated 944 products disclosed by industry. From these products, 632 chemicals were identified, and CAS numbers could be determined for 353 of these. The researchers then classified the biological effects into 12 categories, 1 of which was endocrine disruption. On the basis of their analysis, 37% of these chemicals were determined to have a negative effect on the endocrine system! However, the databases may underrepresent EDC effects because many of these chemicals have not been systematically evaluated and many candidate chemicals remain untested.

Building on Colborn's work, Kassotis and colleagues hypothesized that a selected subset of chemicals used in natural gas drilling operations would exhibit estrogen and androgen receptor activities.[12] Focusing on surface and groundwater in a drilling-dense region of Garfield County, Colorado, the researchers took multiple samples of surface, ground, and artesian well water as well as samples from the Colorado River; the Colorado River is the drainage basin for this region. The researchers also took several control samples, that is, water from sites in Garfield County with a lower density of wells and water from a county without UGE activity (Boone County, Missouri).

Findings showed that the majority of water samples collected from sites in a drilling-dense region of Colorado exhibited more estrogenic, antiestrogenic, or antiandrogenic activities than reference sites with limited nearby drilling operations. The data suggest that natural gas drilling operations may result in elevated endocrine-disrupting chemical activity in surface and groundwater.

Findings from this study should give one pause. The evidence strongly indicates that the chemicals used in UGE, more likely than not, could raise the risk of reproductive, metabolic, neurological, and other diseases, especially in children who are exposed to EDC. Given these findings, it would be prudent to identify the chemicals used in UGE to test the hypothesis that it is effluent from natural gas operations that is the source for these EDCs and not some other sources such as population wastewater or agriculture.

Air pollutants are released during all phases of UGE. For example, at every stage of gas operations, there is the possibility of unintentional gas leaks (fugitive emissions) that can result from small cracks or breaks in seals, tubing, valves, or pipelines or from improperly closed or tightened lids or caps on equipment or tanks. Fugitive emissions release the methane, the primary component of natural gas, as well as VOCs. Areas where there is gas production have reported significant increases in ozone (smog) due to the release of VOCs nitrogen oxides during the UGE process. Smog-causing VOCs, methane, and cancer-causing air emissions are all released during the UGE process. Harmful VOCs typically are not acutely toxic but have compounding long-term health effects.

Air quality studies have been conducted in areas where unconventional drilling is ongoing. Pétron et al. looked at VOC and methane emissions from UGE in the Denver-Julesburg fossil fuel Basin (DJB) in northeastern Colorado,[13,14] and Warneke et al.[15] conducted studies in the Uinta Basin in Utah. Clearly, more research is needed to empirically assess the potential for harm, but the findings themselves are troubling. Chapter 5 by De Jong, Witter, and Adgate in this text provides more information on air quality studies.

In an effort to be somewhat transparent, the energy industry created a website called FracFocus, a voluntary chemical disclosure information source categorizing chemicals arranged geographically for specific unconventional natural gas wells.[16] It is important to stress that the lists of chemicals are not comprehensive because the industry reserves the right to withhold the CAS numbers of chemicals, claiming that such information is proprietary and necessary to ensure their competitive advantage.

It should be noted that even if industry discloses the CAS numbers of all the chemicals used in natural gas operations using databases similar to FracFocus, this does not identify all the potentially active EDCs entering the environment. The tight formations of shale and sand contain many water-soluble chemicals and VOCs that have potential EDC activity. Researchers to date have not published studies systematically identifying these chemicals in the peer-reviewed literature. Those studies released that have measured this chemical composition have used assays that may be insufficiently sensitive to detect all EDCs given that some are active at low concentrations. Also, the tight shale formations themselves differ in the chemical composition of the brines, heavy metals, hydrocarbons, and VOCs. Thus, the presence or concentrations of particular EDCs in the flowback and produced waters likely differ, particularly between and to a lesser extent within shale formations. This situation makes it difficult to generalize epidemiological findings from one area to another *unless the chemical composition of the flowback and produced waters are completely characterized*. As already noted, this information is not available.

EXPOSURE ROUTES WILL VARY

Unlike studying the effects of many industries where there is a single point source of exposure, UGE is distributed over large areas with multiple exposure point sources. Individuals live at different distances from well pads, which will influence their exposure to the harmful effects of the drilling and extraction of natural gas. In addition, the prevailing winds or the flow direction of groundwater is a factor that must be taken into account. Furthermore, individuals living at the same distance from well pads may have different exposure patterns. The kinds of exposure change throughout the drilling and production process. For example, at one specific time point, the exposure may be to truck traffic and chemical spills, whereas at another time point from well stimulation, venting, and flaring. The disposal pathways for wastewater and drilling cuttings vary with geography and may be at some distance from the sites of production, placing a totally different population at risk. Thus, dose exposure, length of time of exposure, and type of exposure must be considered before conclusions can be drawn.

The demographics of the population at risk must also be considered. The distribution of age, socioeconomic status, and ethnicity and the proportion of vulnerable populations (e.g., pregnant women, children, and the elderly) would affect the susceptibility to the health effects of EDCs. Individual EDCs have varying effects on subjects depending on gender, age, and the presence of critical developmental time windows, chemical concentration, and duration of exposure. The effects of mixtures of EDCs or the presence of other drugs or disease processes may also affect the clinical manifestations of EDCs. Some of these health effects may be immediate; others may occur over much longer periods of time (latency or lag time). The manifestations of EDC activity may be only on one particular physiological system—for example, reproduction—or there may be multisystem effects, including those on the immune system and the genome. Some chemicals may have toxic manifestations at one concentration range or with one route of exposure; others may have EDC activity at a different concentration or at a particular critical developmental window (e.g., arsenic and cadmium). All of these concerns must be taken into account when assessing the potential for harm to health and well-being.

SCREENING FOR EDCs

Concerns about chemicals disrupting the endocrine system in humans and animals date from the 1990s. A variety of chemicals were found to disrupt the endocrine systems of animals in laboratory studies, and compelling evidence showed that endocrine systems of certain fish and wildlife have been affected by chemical contaminants, resulting in developmental and reproductive problems. In response, in 1996 Congress passed the Food Quality Protection Act and the Safe Drinking Water Act Amendments that required the Environmental Protection Administration (EPA) to screen pesticide chemicals for their potential to produce effects similar to those produced by the female hormones (estrogen) in humans. These acts gave the EPA the authority to screen certain other chemicals and to include other endocrine effects.[17] In particular, the Food Quality Protection Act added a section to the Federal Food, Drug, and Cosmetic Act that specifically required the EPA to develop a screening program using appropriate validated test systems and other scientifically relevant information to determine whether certain substances may have an effect in humans that is similar to an effect produced by a naturally occurring estrogen, or other such endocrine effect as the administrator may designate.

The EPA created the Endocrine Disruptor Screening Program (EDSP), a massive chemical-testing program intended to screen chemicals for endocrine disruption.[18] The EPA has devised a two-tiered testing program to assess the chemicals: Tier 1 screening assays consist of five in vitro (cell-based) and six in vivo (animal) assays (the Tier 1 battery) and are intended to identify chemicals that may have the potential to cause endocrine disruption. These chemicals would then be further evaluated with additional animal testing in Tier 2, which is designed to confirm the adverse endocrine effect on animals and establish a dose–response relationship.

WHAT CAN EPIDEMIOLOGY OFFER?

The need for epidemiological investigations is long past due. That being said, a large part of the problem is that for a wide range of endocrine disrupting effects, agreed and validated test methods do not exist. In many cases, even scientific research models that could be developed into tests are missing. Until better tests become available, hazard and risk identification have to rely on epidemiological approaches.[19] Descriptive studies would provide some information but would not be able to determine causality. Trends in morbidity and mortality would give a snapshot over time. Databases containing statistics on stillbirths, birth defects, Apgar scores, infant mortality rates, and specific diseases should be evaluated. In smaller, defined populations, blood, urine, or tissue biopsy can be used to evaluate the levels of individual EDCs in the organism or to look for effects related to EDC activity. In particular, assessing hemoglobin A1c, thyroid function tests, or levels of sex hormones would help determine insults to the human body from UGE. Rapid throughput sequencing technology and related molecular techniques may introduce novel ways of looking at epigenetic processes in population studies, but these tests are expensive.

POLICY IMPLICATIONS

As the pace of UGE increases, a serious effort must be made to better understand the relationship between EDCs and health outcomes. EDCs have been shown to have transgenerational and epigenetic effects, so wouldn't it be prudent to have statistical facts now rather than deal with the consequences in future generations? The Endocrine Society Scientific Statement laid out some sensible recommendations for consideration.[20]

- Animal studies should be conducted to help identify which EDCs should be studied in humans. Such data should be used to inform policy decision regarding human exposures to EDCs.
- Studies to identify populations or subgroups with high exposures to EDCs should be undertaken, and exposure–response studies among these populations should be done.
- Epidemiologic studies should incorporate measurement of exposure to multiple EDCs.
- Epidemiologic studies should examine the relationships between EDC exposures (particularly agents with estrogenic and antiandrogenic activity) and measurable endpoints.
- Validated biomarkers of EDC exposures and relevant outcomes should be incorporated into epidemiologic studies.

At present, legislators have little understanding of EDCs and their impact on human health. Yet given what we do know about EDCs, there needs to be an active educational program to inform policymakers of the nature and scope of EDCs and what impact they could have on health and well-being. This is necessary to craft policy recommendations, regulations, and laws. Endocrinologists and scientists researching EDCs should be included in the process of environmental and health impact assessments. They also ought to be included in the scientific panels appointed by policymakers who write advisory reports and who draft legislation.

Policymakers in states with active UGE activity need to legislate for complete transparency for the unique identities, concentrations, and uses for each of the chemicals used. The lack of industry's full disclosure of chemicals used in UGE is a major obstacle. Legislators should pass laws and regulations calling for complete open access for researchers to measure toxicant and EDC concentrations and activities in water and air samples from the entire life cycle of UGE, including access to the flowback, produced waters, wastewater storage and disposal sites and air samples from production gas, vented or flared gas, and fugitive emissions from well casings, compressor stations, and other points in the transmission chain. In addition, policymakers should legislate to provide access to UGE operations and funding for academic researchers and citizen scientists to conduct sound research to evaluate the short- and long-term effects of this industry on public health.

CONCLUSIONS

As many of the chapters in this book discuss, UGE poses numerous environmental and health concerns that are not fully being addressed. These are treated as externalities by the energy industry, which is prompted to act only when forced to do so. As such, it is important for well-designed, empirically based studies to be conducted to provide unbiased evidence from which rational policy can be designed.

Scientists have a key role in providing evidence about the environmental and health effects of these limits. EDCs are only now being recognized as a consequence of UGE operations. Not only can EDCs cause changes in physiological systems, they can also affect growth and development. The epigenetic effects of EDCs must be understood. Unfortunately, as of this writing, research is hampered by an incomplete listing of chemicals released into the environment throughout the UGE process.

The growing body of evidence warrants strong regulations on industry to mandate the release of the identities and concentrations of all the chemicals used in UGE. The huge potential externalized costs on the environment and human health should outweigh any competitive advantage individual companies may gain from keeping this information an industrial secret. Industry should cooperate with scientists who are studying EDCs to allow complete and unconditional access for sampling of gas, water, and air released into the environment from all stages of the life cycle of UGE. Of key concern is how best to mitigate the potential adverse effects of UGE on human health. Understanding the implications of EDCs on human health would be an important and valuable next step.

NOTES

1. Hill, AB. The environment and disease: association or causation? *Proceedings of the Royal Society of Medicine* 58 (1965): 295–300.

2. Rowe BL, Toccalino PL, Moran MJ, et al. Occurrence and potential human-health relevance of volatile organic compounds from domestic wells in the United States. *Environmental Health Perspectives* 115, no. 11 (2007): 1539–1546.

3. Agency for Toxic Substances and Disease Registry. "Exposure Evaluation: Evaluating Exposure Pathways." Chapter 6 of *Public Health Assessment Guidance Manual* (2005 Update). http://www.atsdr.cdc.gov/hac/PHAManual/ch6.html (accessed July 7, 2014).

4. Ibid.

5. Diamanti-Kandarakis G, Bourguignon JP, Giudice L, et al. Endocrine-disrupting chemicals: an Endocrine Society Scientific Statement. *Endocrine Reviews* 30, no. 4 (2009): 293–342. doi: 10.1210/er.2009-0002.

6. State of the science of endocrine disrupting chemicals—2012. An assessment of the state of the science of endocrine disruptors prepared by a group of experts for the United Nations Environment Programme (UNEP) and WHO. http://www.who.int/ceh/publications/endocrine/en (accessed July 7, 2014).

7. Diamanti-Kandarakis et al., Endocrine-disrupting chemicals, op. cit.

8. Hotchkiss AK, Rider CV, Blystone CR, et al. Fifteen years after "Wingspread"—environmental endocrine disrupters and human and wildlife health: where we are today and where we need to go. *Toxicology Science* 105, no. 2 (2008): 235–259.

9. Diamanti-Kandarakis et al., Endocrine-disrupting chemicals, op. cit.

10. Ibid.

11. Colborn T. *Spreadsheet of products, chemicals and their health effects.* TEDX, The Endocrine Disruption Exchange. http://www.endocrinedisruption.com/chemicals.multistate.php (accessed July 7, 2014).

12. Kassotis CD, Tillitt DE, Davis JW, et al. Estrogen and androgen receptor activities of hydraulic fracturing chemicals and surface and ground water in a drilling–dense region. *Endocrinology* 155, no. 3 (2014). DOI: http://dx.doi.org/10.1210/en.2013-1697. http://press.endocrine.org/doi/full/10.1210/en.2013-1697 (accessed November 6, 2014).

13. Pétron G, Frost GJ, Trainer MK, et al. Hydrocarbon emissions characterization in the Colorado Front Range. A pilot study. *Journal of Geophysical Research* 117 (2012): D04304. doi:10.1029/2011JD016360.

14. Pétron G, Frost GJ, Miller BJ, et al. *Estimation of emissions from oil and natural gas operations in northeastern Colorado.* Boulder, CO: Earth System Research Laboratory, National Oceanic & Atmospheric Administration, 2012.

15. Warneke C, Geiger F, Edwards PM, et al. Volatile organic compound emissions from the oil and natural gas industry in the Uinta Basin, Utah: point sources compared to ambient air composition. *Atmospheric Chemistry and Physics Discussions* (2014): 11895–11927.

16. FracFocus 2.0. http://fracfocus.org (accessed July 7, 2014).

17. U.S. Environmental Protection Agency. *Endocrine Disrupting Screening Program.* http://www.epa.gov/endo (accessed July 7, 2014).

18. Ibid.

19. Kortenkamp, A, Martin, O, Faust, M, et al. *State of the Art Assessment of Endocrine Disruptors. Final Report.* http://ec.europa.eu/environment/chemicals/endocrine/pdf/sota_edc_final_report.pdf (date accessed July 7, 2014).

20. Diamanti-Kandarakis et al., Endocrine-disrupting chemicals, op. cit.

Chapter 3

Impacts of Shale Gas Extraction on Animal Health and Implications for Food Safety*

Michelle Bamberger and Robert E. Oswald

INTRODUCTION

In recent years, the energy industry has extended the extraction of hydrocarbons to increasingly challenging formations. In particular, the extraction of shale gas and tight oil using horizontal drilling with high volume hydraulic fracturing is now possible across huge tracts of land throughout the United States and the world.[1] Although this technology has created a boom in U.S. energy production, there are negative consequences that must be addressed. Industrialization of the landscape has occurred in areas that were once largely dedicated to farming, tourism, and outdoor recreation, as well as to residential areas. Unconventional extraction of oil and gas has brought an influx of workers to predominantly rural

---

* The opinions expressed in this communication are the authors' own and do not represent the views of Cornell University in any way. Nothing written here should be considered official or sanctioned by Cornell University or any other organization with which they are affiliated.

communities as well as heavy truck traffic on roads unable to withstand such vehicles. Most alarmingly, however, unconventional operations (extraction, processing, and production) introduce large quantities of toxic chemicals into these communities along with the potential for contamination of water, air, and the food supply. Such chemicals include components of drilling and fracturing fluids, substances extracted from ancient shale layers (organic compounds, heavy metals, radioactive substances, bacteria, archaea, etc.) and airborne pollutants such as ozone and volatile organic compounds (VOCs). Although extraction of oil from tar sands (also referred to as oil sands) in Canada[2] and the planned expansion of tar sands extraction in Utah and Colorado[3] —which may eventually dwarf the environmental devastation in Alberta—suffer from many of the same problems described in this book, we will concentrate only on the health issues surrounding shale gas and tight oil extraction.

There is a paucity of research on the health concerns surrounding unconventional oil and gas development on humans, and there is little to no research on animals. Of the few studies that have been done, most focus on wildlife; the literature on the health impacts to both farm animals and companion animals is scant.[4] This is unfortunate for many reasons, not the least of which is that animals can be viewed as sentinels of human health, so that if we understand the environmental toxicology associated with animal health, we can extrapolate what may be in store for humans. Perhaps most important, however, is that production animals, such as cattle and chickens, constitute a vital part of the food chain, and their toxic load becomes our toxic load. The implications for food safety should be of great concern yet have largely been ignored.

This chapter focuses on the potential for harm to food animals, companion animals, and wildlife and the implications for food safety. In cases in which there is a dearth of research involving unconventional industrial operations, we also discuss health impacts associated with conventional oil and gas drilling, processing, and production. Although few definitive answers are currently available, new and innovative testing methods are coming online that provide hope to better understand these issues in more detail in the near future.

FOOD ANIMALS

Food safety has arisen as a major concern relative to unconventional oil and gas extraction, and for good reason: well sites are drilled in the middle of cornfields or nearby ponds or streams that serve as sources of water for cattle; and pipelines, processing plants, and compressor stations are often surrounded by grazing cattle and deer. In addition, the practice of land farming[5,6] (a process in which drilling waste or wastewater is disposed on farmland, thus introducing toxic chemicals including radioactive compounds[7] into the soil) has received only limited attention. Although the impact of land farming remains in question, some food producers are taking no chances and are refusing milk from dairies engaged in this practice due to the high cost of testing for contamination.[8]

No one would argue that clean air and water are required to produce healthy, safe food, but to date, there has been no systematic study of the potential contamination of our food supply by chemicals associated with unconventional oil and gas operations. Furthermore, no systematic testing of food that would detect contamination due to by-products of the fossil fuel industry is required. Slaughterhouses do not systematically or randomly test food animals for chemical contamination in the United States. Cows, for example, are culled from the food supply for gross lesions and if they are unable to walk; however, their tissues, along with those of swine and poultry, still gain entrance into our food supply through the rendering process.[9] This lack of testing at both the federal and state levels does not imply that no harm is being done.

Although there are currently few published reports on food animal exposures to contaminants from unconventional shale gas extraction,[10] the literature does contain several reports on exposure of sheep[11] and cattle[12–19] to conventional oil and gas operations. In addition, changes in milk production and herd size in intensively drilled areas in Pennsylvania have been noted.[20]

More than 20 years ago Adler et al.[21] reported on pathological findings in sheep after a 1-day exposure to natural gas condensate resulting from a valve leak on a storage tank. Exposed ewes either died or were euthanized over a 3-week period postexposure; clinical signs included dyspnea, recumbency, rumen atony, bloody diarrhea, and sudden death. The main finding on necropsy was aspiration pneumonia, although many other organ systems were affected. Aspiration of chemicals typically follows ingestion-induced eructation or emesis in ruminants, leading to pneumonia.[22,23] Because controls were not available and dose–response studies could not be done, toxicological analysis was used to confirm causality. The gas chromatogram of the condensate matched the chromatograms of the rumen contents (a process known as fingerprinting) and contained mostly cyclic aromatic hydrocarbons including benzene, which is a known carcinogen and immunotoxicant.[24]

Early studies on beef cattle also raise concern. Waldner and collaborators, building on findings from a previous report[25] that found associations between increasing exposure to sour-gas flares and calf mortality and between sour-gas flares and increased risk of stillbirth, initiated a detailed series of studies on the effects of emissions from nearby conventional oil and gas operations on beef cattle reproduction and health. Herds were recruited from areas of western Canada and were grouped by exposure level based on distances to well and battery sites (production facilities), compressor stations, and processing plants. Random herd selection would have produced a more accurate analysis, but there was no complete listing of all cow-calf operations that also included herd size and pasture locations in western Canada at the time these studies were conducted. Passive air monitors located in pasture, wintering, or calving areas were used to measure monthly concentrations of sulfur dioxide, hydrogen sulfide, and VOCs.

Looking at the association between exposure to byproducts of the oil and gas industry and beef calf mortality, Waldner.[26] found that exposure of cows to sulfur dioxide during the past 3 months of gestation was associated with an increased risk of calf mortality and that there was an increased occurrence of respiratory lesions in liveborn calves associated with increasing exposures to VOCs measured as benzene and toluene. In this same report, the researchers also found that with increasing exposure to sulfur dioxide, there was an increased occurrence of degeneration and necrosis in the skeletal and heart muscle of necropsied calves. In a subsequent report, Waldner and Clark[27] investigated associations between postnatal exposures to oil and gas facility emissions and the risk of lesions to the respiratory, nervous, immune, and muscular systems of beef calves. They found that with increasing age of the calf, there was an increased risk of developing respiratory lesions associated with increasing exposures to benzene.[28]

Bechtel et al.[29] also studied the immune systems of neonatal calves (1–7 days of age) born to exposed cows to evaluate potential immunosuppressant effects of emissions from oil and gas facilities. The researchers determined that CD4 T-lymphocyte counts were significantly reduced when associated with increasing VOCs measured as benzene and toluene and that likewise, CD8 T-lymphocyte counts were significantly reduced when associated with increasing VOCs measured as benzene.

To investigate the possibility of immunomodulation due to chronic exposure to emissions from oil and gas facilities, Bechtel et al.[30] studied the exposure of yearling beef cattle to airborne sulfur dioxide, hydrogen sulfide, and VOCs by measuring lymphocyte numbers in peripheral blood and measuring antibody production following vaccine challenge. They found that CD4 T-lymphocyte counts were significantly reduced when associated with increasing VOCs measured as toluene.

These studies provide a comprehensive look at a large number of animals using methods that, at the time, were state of the art. Statistically significant associations with chemical exposures and biological outcomes were measured and carefully reported. These findings raise important questions that not only illustrate the need for further study but also provide important and interesting clues as to the next steps that should be taken. In an editorial, the editor of the journal in which these studies were published, who was also the cochair of the scientific advisory panel for the study, dismissed the significance of the findings: "With specific anomalies and exceptions noted in the report, there was no overall pattern to suggest substantial adverse effects on cattle reproduction or wildlife."[31] Without further study, we will never know the significance of these "anomalies and exceptions."

In a study that we conducted in six states in the United States, we used descriptive epidemiology in the form of 24 case reports as a first step in studying the health impacts of unconventional gas extraction on food animals, companion animals, wildlife, and animal owners.[32] This study was not designed to understand the prevalence of a problem or to demonstrate cause and effect. Instead, the purpose was to generate hypotheses that should be empirically tested in analytical studies. Information was collected on drilling and production operations, environmental test results, sources of exposure, and health records of animals and their owners living in locations of intensive drilling activity. Farm animals consisted of beef and dairy cattle, goats, and chickens from 12 farms; cattle herd sizes ranged from 20 to 100 head. Most of the exposures to farm animals were directly related to wastewater, drilling muds, or hydraulic fracturing fluid. On several farms, herds were divided on different pastures when exposures occurred, some in close proximity to drilling activity and others not, thus allowing a natural control. Findings showed that a proportion of the exposed cattle suffered reproductive, neurological, and growth problems, and the unexposed cattle experienced no unusual health problems. On these farms, neurological problems occurred within 48 hours following exposure, whereas reproductive and growth problems appeared months later.

We determined that, overall, animals and their owners living in intensively drilled areas often suffered from a similar set of acute symptoms that has been dubbed "shale gas syndrome" in which the neurological, dermatological, gastrointestinal, respiratory, and vascular systems are affected. However, the most prevalent major health impact experienced by farm animals and companion animals was in reproduction, for example, failure to breed, failure to cycle, abortions, and stillbirths.[33] With time, reproductive problems may begin to appear in humans as well, but we are most likely seeing these problems now in animal sentinels because of longer exposures and shorter generation times. Long-term follow-up of farm animals (more than 2 years after initial interviews) showed that reproductive problems have decreased but that respiratory problems (coughing, wheezing, heaving, difficulty breathing) and growth problems (stunting and failure to thrive) increased dramatically.[34]

Although farmers typically keep careful records of their operations, veterinary health records for farm animals are not kept in centralized databases, so that detailed quantitative epidemiology on farm animals is difficult. However, statistics on milk production, number of cows, and production per cow are available in Pennsylvania, although they are updated only periodically (e.g., 5-year cycle for the U.S. Department of Agriculture's Census of Agriculture) and are reported as totals for a particular county rather than for individual farms. Finkel et al.[35] analyzed milk production and the number of milk cows between 1996 and 2011 in Pennsylvania counties with active wells and neighboring counties with few or no wells. Findings showed that regardless of the number of active wells, both milk production and number of cows had decreased in most Pennsylvania counties since 1996. However, with the wide-scale onset of unconventional gas drilling in the mid-2000s, greater decreases occurred from 2007 through 2011 in five counties with the most wells drilled compared with six adjacent counties (fewer than 100 wells drilled). Similar results in Pennsylvania were reported by Adams and Kelsey.[36] These studies do not provide answers as to why milk production and the number of milk cows decreased in counties with intensive drilling activity; however, they do show a consistent drop in production. This finding is significant in that Pennsylvania is a top-ranked dairy state.[37]

In light of the results discussed here and the worldwide increase in unconventional operations, further studies of the possible link among toxic exposures, biological outcomes, animal health, and intensive shale gas and tight oil extraction are required. Although much has been made of the industry-supported after-the-fact release of some of the chemicals used in hydraulic fracturing,[38] the fact remains that we cannot at this point definitively state which possible toxicant or combination of toxicants may be present in the environment near oil and gas operations. Also, we cannot be certain what routes of exposure may be the most important, although air and water contamination are high on the list. Furthermore, the levels of toxicants will never be constant so that single samples of air, water, or soil cannot tell the whole story.

Even if we detect and identify specific environmental toxicants, the interpretation of the results remains difficult for several reasons. Maximum contaminant levels have not been determined for the majority of toxicants associated with unconventional operations. Second, low dose effects (i.e., effects on biological systems that are significant but below concentrations that have previously been defined as toxic) as well as the length of time exposed are known to be important factors to consider, but more research must be done.[39] Finally, the effects of multiple toxicants cannot be easily assessed with the available data.

Fortunately, some innovative approaches are beginning to be applied to this problem. Kassotis et al.[40] used a biological assay (effects on estrogen and androgen receptors) to detect the presence of endocrine disrupting chemicals in water near gas drilling operations. This approach does not ask the question of what is in the water; rather, it asks simply whether something is in the water that has a significant biological effect (in this case, interaction with estrogen and androgen receptors). This is an important approach that complements perfectly high-resolution chemical methods.

In this regard, Allen et al.[41,42] have used passive absorption devices to collect a wide range of organic compounds from air and water over an extended period of time (up to several weeks). This circumvents the problem of single-grab samples in that it is a time-integrated sample. In such a sampling procedure, any organic compounds present in the environment (air or water) during the sampling period can adhere to the sampling device and subsequently be detected. In that way, the procedure is much less susceptible to temporal variations in toxicant levels. The compounds absorbed by the devices are extracted and subjected to a screen of approximately 1,200 compounds using tandem gas chromatography–mass spectrometry. These new approaches offer the promise of finally understanding the toxicant load of food producing animals and people, which will be important not only for the farmer but also for the consumer purchasing food products from areas where active oil and gas operations are ongoing.

COMPANION ANIMALS

Little attention has been given to the possibility of adverse health effects on companion animals due to unconventional oil and gas extraction. We define companion animals here as any animals living with humans and not used for food production. Cats, dogs, birds, horses, and other species bring pleasure and provide valuable services to humans such that their health is of paramount importance: Americans spent almost $56 billion on their pets in 2013, and this number was estimated to rise in 2014.[43] However, they also provide another often forgotten but invaluable service: that of sentinels of environmental disease. When children and adults leave for school and work, their pets are often left behind in the house, yard, or barn. This increases the animal's exposure time compared with their human counterparts and is one of the reasons pets are often the first in the household to become ill.

Our case study described earlier[44] also included case reports of companion animals (horses, dogs, cats, llamas, goats, and koi) living nearby unconventional shale gas operations. Findings showed that in companion animals, ingestion of contaminated well or spring water accounted for most of the exposures. As with food animals, reproductive problems were the most commonly reported health concerns, although neurological problems (seizures, incoordination), gastrointestinal (vomiting, diarrhea), and dermatological (hair and feather loss, rashes) problems were also reported. Long-term follow-up on these companion animal cases has shown no health differences over time; however, in all cases where owners moved to locations of little to no shale gas industrial activity or where the level of industrial activity decreased in the neighborhood, the health of the companion animals (as well as the health of their owners) dramatically improved.[45]

Slizovskiy et al.[46] used a survey to study animal health near shale gas extraction activity. They reported a spatial correlation between dermatological complaints in companion animals and areas having the most dense natural gas extraction activity in Washington County, Pennsylvania. As unconventional shale gas activity expands into neighborhoods, it is likely that there will be more negative health effects on companion animals. More study is needed to clarify cause and effect via further epidemiological and empirical analyses.

WILDLIFE

The effects of unconventional gas extraction on wildlife have been studied more extensively. Wildlife bestow both recreation and food to humans, function to maintain ecosystems, and simply have a right to exist without excessive human interference. In Pennsylvania,[47] and likely many other states permitting unconventional oil and gas extraction, the majority of well pads occur on agricultural lands and on privately held core forest; both are prime habitat areas where wildlife live and reproduce. Within these forested areas, the spatial footprint of unconventional oil and gas drilling can be large, including well pads, disturbance to the land via access roads and associated drilling mud pits, wastewater impoundment sites, storage sites, compressor stations, processing plants, and pipelines. Drilling muds pits have been reported to entrap migratory birds and other wildlife,[48] and we have documented several cases in which both foxes and deer have become entrapped in wastewater impoundments in Pennsylvania.[49]

Shale gas development also causes forest fragmentation—not only because of well pads and the access roads to the well pads but also because of the associated infrastructure—in particular, clear cuts for pipeline right-of-way. Because headwater streams often are found in forests where there is drilling activity, there is a real risk of pollution of these streams and waterways located further downstream due to run off from well pads, as well as leaks of impoundments and tanks holding drilling muds, drilling and hydraulic fracturing fluid, and wastewater.[50] Always present is an increase in noise pollution due to truck traffic, flaring, and compressor station and processing plant operations. Francis et al.[51] analyzed the effects of chronic gas well compressor noise on two avian species in Colorado (the gray flycatcher, *Empidonax wrightii*, and the western scrub-jay, *Aphelocoma californica*). The occupancy of both species was found to be lower than would be expected without compressor noise. These researchers also discovered that the nesting success of the gray flycatcher increased in noisy areas due to a decreased rate of predation by the western scrub-jay on the gray flycatcher. The authors emphasize that although only two species were studied for this work, many more avian species may be noise sensitive. Also, other ecological parameters may be influenced, such as decreased clutch size and pairing success. To attempt to mitigate these effects on wildlife by returning to normal baseline noise levels, the researchers advise that walls be placed around individual compressors as well as compressor stations.

Because of the practice of land spreading (legal dumping of drilling fluids, muds, hydraulic fracturing fluids or wastewater on forest or agricultural land), some information on how these fluids may affect a forest has been reported. When fracturing fluids were applied to a mixed hardwood research forest in West Virginia, immediate damage to ground vegetation occurred followed soon after by loss of foliage, high soil levels of sodium and chloride, and loss of more than half of the trees within 2 years.[52] The impacts on vegetation were easily observed in this study, but impacts on animal species living in this forest were not addressed.

However, in a case in which hydraulic fracturing fluids were illegally dumped into a headwater stream in Kentucky, impacts on animal life were documented. In this case, researchers were informed of this event soon after it happened and were able to study changes in water quality and fish morbidity and mortality.[53] Compared with unaffected water upstream, water quality downstream of this incident was much higher in conductivity, lower in pH, and contained toxic levels of aluminum and iron. These changes in water quality resulted in stressed aquatic invertebrates and fish and in fish with gill lesions; fish also bioaccumulated aluminum and iron. Because the creek is a spawning ground for the threatened Blackside dace, found in the Cumberland River drainage (Kentucky and Tennessee) as well as the Powell River drainage (Virginia), and because the detrimental effects of this incident extended over several months, the authors warn of a long-standing disturbance to the ecology of this stream as well as waters downstream.

In the western United States, where unconventional fossil fuel extraction has been occurring since the late 1990s, health effects of shale gas operations on several species of wildlife have been studied. In Wyoming, Holloran[54] observed that the greater sage-grouse tends to avoid areas undergoing gas development. Numbers of breeding males declined in leks (male display areas) that were closer to gas operations, and nesting females avoided active drilling rigs and producing wells. In Montana and Wyoming, where intensive development of coal-bed methane has changed the winter habitat of the greater sage-grouse, Doherty et al.[55] found that sage-grouse were more likely to choose wintering grounds that lacked wells compared with suitable habitats nearby wells.

In addition to sage-grouse, mule deer habitat selection in intensively drilled areas has also been studied. Using GPS radio collars, Sawyer et al.[56] studied the winter habitat choices of mule deer before and during unconventional development of a natural gas field in western Wyoming. In addition to studying the direct loss of habitat through construction of well pads and access roads, the researchers were also interested in determining the impact of indirect habitat loss caused by increases in human-related noise and associated infrastructure build-out. They discovered that mule deer avoided gas well pad development over both the short term (first year of gas field development) and long term (through 3 years of development) and are concerned that these changes in habitat selection may have an impact on the reproduction and survival of mule deer.

In further studies on mule deer habitat selection in western Wyoming, Sawyer et al.[57] used infrared sensors to monitor vehicle traffic at well pads actively undergoing horizontal drilling and well pads that were in the production phase with a liquids gathering system (LGS; condensate and produced water) and without LGS. The researchers found that in general, mule deer tended to avoid all types of well pads, especially those with higher levels of traffic. They discovered that active drilling pads were associated with much higher levels of indirect habitat loss than were producing well pads but cautioned that while disturbances at active drilling sites are usually short term (6 months to 2 years), producing wells may cause disturbance for decades. When condensate and produced water were collected in pipelines via an LGS instead of being stored on site and removed via trucks, they found that the loss of indirect habitat was reduced. These results indicated that the impacts of industrial gas operations on mule deer may be lessened by minimizing traffic more than 50%.

CONCLUSION

Animals living near areas of unconventional oil and gas operations can be thought of as sentinels for human health impacts. The available research on domestic and companion animals indicates that animals living nearby unconventional oil and gas operations suffer from a set of acute symptoms that are similar to their human counterparts. These symptoms are known collectively as shale gas syndrome and commonly affect the respiratory, gastrointestinal, dermatological, neurological, and vascular systems—and in particular, the reproductive system. Reproductive problems may yet appear in humans, but these symptoms (abortions, stillbirth, failure to breed and cycle) are likely appearing acutely in these animal sentinels because of longer exposures and shorter generation times. Research on wildlife has concentrated on habitat choices and indicates that animals tend to avoid living, breeding, and nesting in areas near unconventional operations, likely because of the noise and traffic associated with drilling operations.

Although there are few definitive answers, many of the studies discussed in this review raise serious issues that need further investigation. The problem, however, is the difficulty of studying health impacts of essentially unknown chemicals and mixtures of unknown chemicals that vary with time through multiple routes of exposure. It is well known that drilling and fracturing fluids, wastewater, and air contaminants released during unconventional operations contain toxic chemicals including carcinogens. Because industrial operations are often located on agricultural land, crops from exposed fields; milk, meat, and eggs from exposed animals; and fish from exposed waterways may be contaminated and should not be made available for human consumption without careful monitoring.

Perhaps the most difficult questions for researchers studying health problems in intensively drilled areas concern testing. A number of innovative methods have been introduced recently (e.g., passive sampling devices coupled to gas chromatography/mass spectrometry and biological assays of endocrine disrupting chemicals) that have the promise to provide some answers. However, this work is being done in a highly politically charged atmosphere, where science is often viewed through the lens of preconceived notions, not to mention financial motivations. In the coming years, it is essential to approach these topics with scientific rigor and innovative methods because human and animal health as well as the safety of our food supply demand such action.

Given the potential for harm, public health initiatives should be proactive, and there is precedence for such action. In the 1990s, federal and state governments acted together to prevent an outbreak of bovine spongiform encephalopathy ("mad cow disease") and variant Creutzfeldt-Jakob disease in the United States.[58] By taking proactive, preventive action before conclusive testing was completed, the public's health was not compromised. Now is the time to make another careful analysis of risk, albeit with incomplete data, to assess the short- and long-term impacts of unconventional fossil fuel extraction on human and animal health.

NOTES

1. Mohr SH, Evans GM. Long term forecasting of natural gas production. *Energy Policy* 39 (2011): 5550–5560.

2. Nikiforuk A. *Tar Sands: Dirty Oil and the Future of a Continent*. Vancouver, Canada: Greystone Books, 2010.

3. U.S. Department of the Interior. Oil shale research, development, & demonstration (RD&D) leases. 2013. http://www.blm.gov/co/st/en/fo/wrfo/WRFO_Oil_Shale_Program.html (accessed March 19, 2014).

4. Bamberger M, Oswald RE. Impacts of gas drilling on human and animal health. *New Solutions* 22 (2012): 51–77.

5. Adams MB. Land application of hydrofracturing fluids damages a deciduous forest stand in West Virginia. *Journal of Environmental Quality* 40 (2011): 1340–1344.

6. Drilling Waste Management Information System. Fact Sheet—Land Application. http://web.ead.anl.gov/dwm/techdesc/land/index.cfm (accessed April 29, 2014).

7. Enviornmental Protection Agency. *Radioactive Wastes from Oil and Gas Drilling*. 2012. http://www.epa.gov/radtown/drilling-waste.html (accessed May 16, 2014).

8. Radio New Zealand News. Fonterra to stop taking milk from farms with oil and gas waste. 2013. http://www.radionz.co.nz/news/rural/138025/fonterra-to-stop-taking-milk-from-farms-with-oil-and-gas-waste (accessed May 16, 2014).

9. National Renders Association. The rendering process. 2014. http://www.nationalrenderers.org/about/process (accessed May 16, 2014).

10. Bamberger and Oswald, Impacts of gas drilling, op. cit.

11. Adler R, Boermans HJ, Moulton JE, et al. Toxicosis in sheep following ingestion of natural gas condensate. *Veterinary Pathology* 29 (1992): 11–20.

12. Bechtel DG, Waldner CL, Wickstrom M. Associations between in utero exposure to airborne emissions from oil and gas production and processing facilities and immune system outcomes in neonatal beef calves. *Archives of Environmental and Occupational Health* 64 (2009): 59–71.

13. Bechtel DG, Waldner CL, Wickstrom M. Associations between immune function in yearling beef cattle and airborne emissions of sulfur dioxide, hydrogen sulfide, and VOCs from oil and natural gas facilities. *Archives of Environmental and Occupational Health* 64 (2009): 73–86.

14. Waldner CL. The association between exposure to the oil and gas industry and beef calf mortality in Western Canada. *Archives of Environmental and Occupational Health* 63 (2008): 220–240.

15. Waldner CL. Western Canada study of animal health effects associated with exposure to emissions from oil and natural gas field facilities. Study design and data collection I. Herd performance records and management. *Archives of Environmental and Occupational Health* 63 (2008): 167–184.

16. Waldner CL, Clark EG. Association between exposure to emissions from the oil and gas industry and pathology of the immune, nervous, and respiratory systems, and skeletal and cardiac muscle in beef calves. *Archives of Environmental and Occupational Health* 64 (2009): 6–27.

17. Waldner CL, Kennedy RI. Associations between health and productivity in cow-calf beef herds and persistent infection with bovine viral diarrhea virus, antibodies against bovine viral diarrhea virus, or antibodies against infectious bovine rhinotracheitis virus in calves. *American Journal of Veterinary Research* 69 (2008): 916–927.

18. Waldner CL, Ribble CS, Janzen ED. Evaluation of the impact of a natural gas leak from a pipeline on productivity of beef cattle. *Journal of the American Veterinary Medicine Association* 212 (1998): 41–48.

19. Waldner CL, Ribble CS, Janzen ED, et al. Associations between oil- and gas-well sites, processing facilities, flaring, and beef cattle reproduction and calf mortality in western Canada. *Preventive Veterinary Medicine* 50 (2001): 1–17.

20. Finkel ML, Selegean J, Hays J, et al. Marcellus Shale drilling's impact on the dairy industry in Pennsylvania: a descriptive report. *New Solutions* 23 (2013): 189–201.

21. Adler et al., Toxicosis in sheep, op. cit.

22. Ibid.

23. Coppock RW, Mostrom MS, Stair EL, et al. Toxicopathology of oilfield poisoning in cattle: a review. *Veterinary and Human Toxicology* 38 (1996): 36–42.

24. Agency for Toxic Substances and Disease Registry. Toxicological profile for benzene. 2007. http://www.atsdr.cdc.gov/toxprofiles/tp3-p.pdf (accessed May 16, 2014).

25. Waldner et al., Evaluation of the impact of a natural gas leak, op. cit.

26. Waldner CL. The association between exposure to the oil and gas industry and beef calf mortality, op cit.

27. Waldner CL, Clark EG. Association between exposure to emissions from the oil and gas industry, op. cit.

28. Ibid.

29. Bechtel et al., Associations between in utero exposure to airborne emissions. Op. cit.

30. Bechtel et al., Associations between immune function in yearling beef cattle and airborne emissions, op. cit.

31. Guidotti TL. The Western Canada study: effective management of a high-profile study of risk. *Archives of Environmental and Occupational Health* 64 (2009): 3–5.

32. Bamberger and Oswald. Impacts of gas drilling on human and animal health, op. cit.

33. Ibid.

34. Bamberger M, Oswald RE. Long-term impacts of unconventional drilling operations on human and animal health. *Environmental Science and Health Part A* 50, no. 5 (2015): 10.

35. Finkel et al, Marcellus Shale drilling's impact, op. cit.

36. Adams R, Kelsey TW. *Pennsylvania dairy farms and Marcellus Shale, 2007–2010*. 2012. http://pubs.cas.psu.edu/FreePubs/PDFs/ee0020.pdf (accessed May 16, 2014).

37. California leads U.S. milk production; Ohio dairy herd numbers drop. *Farm and Dairy*. February 22, 2014. http://www.farmanddairy.com/news/california-leads-u-s-milk-production-ohio-dairy-herd-numbers-drop/177273.html (accessed May 16, 2014).

38. Chemical Disclosure Registry. FracFocus 2.0. 2014. http://www.fracfocus.org (accessed May 16, 2014).

39. Vandenberg LN, Colborn T, Hayes TB, et al. Hormones and endocrine-disrupting chemicals: low-dose effects and nonmonotonic dose responses. *Endocrine Reviews* 33 (2012): 378–455.

40. Kassotis CD, Tillitt DE, Davis JW, et al. Estrogen and androgen receptor activities of hydraulic fracturing chemicals and surface and ground water in a drilling-dense region. *Endocrinology* 155 (2014): 897–907.

41. Allan SE, Smith BW, Anderson KA. Impact of the deepwater horizon oil spill on bioavailable polycyclic aromatic hydrocarbons in Gulf of Mexico coastal waters. *Environmental Science and Technology* 46 (2012): 2033–2039.

42. Allan SE, Sower GJ, Anderson KA. Estimating risk at a Superfund site using passive sampling devices as biological surrogates in human health risk models. *Chemosphere* 85 (2011): 920–927.

43. American Pet Products Association. Pet industry market size & ownership statistics. 2014. http://www.americanpetproducts.org/press_industrytrends.asp (accessed May 16, 2014).

44. Bamberger and Oswald. Impacts of gas drilling on human and animal health, op. cit.

45. Bamberger and Oswald. Long-term impacts of unconventional drilling operations on human and animal health, op. cit.

46. Slizovskiy IB, Lamers VT, Trufan SJ, et al. Animal sentinel health surveillance near shale gas extraction activity: challenges and possibilities. *Environmental Science and Health Part A*, forthcoming 2015.

47. Drohan PJ, Brittingham M, Bishop J, et al. Early trends in landcover change and forest fragmentation due to shale-gas development in Pennsylvania: a potential outcome for the Northcentral Appalachians. Environmental Management 49 (2012): 1061–1075.

48. Ramirez P. U.S. Fish & Wildlife Service, Region 6, Environmental Contaminants Program. Reserve pit management: risks to migratory birds. 2009. http://www.fws.gov/mountain-prairie/contaminants/documents/ReservePits.pdf (accessed July 14, 2011).

49. Bamberger M, Oswald RE. *The Real Cost of Fracking: How American's Shale Gas Boom Is Threatening Our Families, Pets, and Food*. Boston: Beacon Press, 2014.

50. Drohan PJ et al., Early trends in landcover change and forest fragmentation, op cit.

51. Francis CD, Paritsis J, Ortega CP, et al. Landscape patterns of avian habitat use and nest success are affected by chronic gas well compressor noise. *Landscape Ecology* 26 (2011): 1269–1280.

52. Adams, Land application of hydrofracturing fluids, op cit.

53. Papoulias DM, Velasco AL. Histopathological analysis of fish from Acorn Fork Creek, Kentucky, exposed to hydraulic fracturing fluid releases. *Southeastern Naturalist* 12 (2013): 91–111.

54. Holloran MJ. *Greater sage-grouse* (Centrocercus urophasianus) *population response to natural gas field development in western Wyoming*. Laramie: University of Wyoming, 2005.

55. Doherty KE, Naugle DE, Walker BL, et al. Greater sage-grouse winter habitat selection and energy development. *Journal of Wildlife Management* 72 (2008): 187–195.

56. Sawyer HR, Nielson RM, Lindzey F, et al. Winter habitat selection of mule deer before and during development of a natural gas field. *Journal of Wildlife Management* 70 (2006): 396–403.

57. Sawyer H, Kauffman MJ, Neilson RM. Influence of well pad activity on winter habitat selection patterns of mule deer. *Journal of Wildlife Management* 73 (2009): 1052–1061.

58. Hueston WD. BSE and variant CJD: emerging science, public pressure and the vagaries of policy-making. *Preventive Veterinary Medicine* 109 (2013): 179–184.

Chapter 4

The Importance of Health Impact Assessments

Liz Green

INTRODUCTION

Shale gas fracturing (hydraulic fracturing, unconventional gas extraction [UGE], or "fracking") is acknowledged internationally as a controversial process. Many governments, including the United States, have embraced shale gas as a plentiful energy source and resource, but others (e.g., France and Bulgaria) have banned UGE, expressing environmental, health, and well-being concerns about the process. Although advances in technology have enabled the economical extraction of shale gas, there have been few well-designed health impact assessment (HIA) studies. An HIA study is designed to provide needed information about potential health impacts of a process. This chapter describes what HIAs are; outlines their principles, methods, and process; and discusses the advantages and challenges in the context of UGE development.

WHAT IS AN HIA?

HIA is defined as a combination of procedures, methods, and tools through which a policy, program, or project may be judged as to its potential effects on the health of a population, and the distribution of those effects within the population.[1] This definition, known as the Gothenburg Consensus, was developed by a group of HIA practitioners in 1999. The Gothenburg Consensus Paper outlines the main concepts and suggested approach to HIA and is identified as a first step to creating a common understanding of HIA.

A major objective or purpose of an HIA is to inform and influence decision making; however, it is not a decision-making tool per se. HIA is a process that considers to what extent the health and well-being of a population may be affected by a proposed action, be it a policy, program, plan, or project. It provides a systematic, objective, yet flexible and practical, way of assessing potential positive and negative health impacts associated with a particular activity. It also provides an opportunity to suggest ways in which health risks can be minimized and health benefits maximized.

In most uses of HIA, "health" is viewed as holistic and encompasses mental, physical, and social well-being. On the basis of a social determinants framework, HIA recognizes that there are many, often interrelated factors that influence people's health, from personal attributes and individual lifestyle factors to socioeconomic, cultural and environmental considerations.

Although some impacts on health determinants may be direct, obvious, and/or intentional, others may be indirect, difficult to identify, and unintentional. An HIA can identify health inequalities not only in the general population but also in "vulnerable groups" as well (e.g., children, young people, or older individuals). That being said, the main output of any HIA is an evidence-based set of recommendations that should lead to the minimization of risks and maximization of potential benefits. It can provide opportunities for health improvement.

THE PRINCIPLES OF HIA

HIA methodology evolved from environmental impact assessment (EIA) tools used in the late 1970s and early 1980s. (See the Merseyside Guidelines,[2] acknowledged to be the first comprehensive guidance on EIAs.) In the context of shale gas development, for example, human health risk can be assessed within an EIA designed to assess potential environmental impacts of a proposed plan or action. However, most EIAs do not take into account the wider determinants of health. Wales in the United Kingdom is an exception. In Wales, all open cast mining development planning applications must be accompanied by an HIA carried out as part of the statutory EIA.[3]

The principles of HIA are made explicit in the Gothenburg Consensus.[4] Briefly, the process should be *open* and *transparent*; the use of evidence and methods of participation should be *ethical* and *robust*; a wide range of stakeholders should be involved to ensure the process is *participatory* and *democratic*; impacts that are short and long term, direct and indirect should be considered to inform policies and projects that are *sustainable*; and there should be a presumption in favor of reducing health inequalities so that the process is *equitable*. Although many of these principles can be difficult to achieve, HIA does emphasize the importance of framing a general agreement of the meaning of these principles from the outset.

Over the decades, many guides and tools have been developed to support policymakers and practitioners; for example, International Association of Impact Assessment (IAIA),[5] HIA Gateway,[6] the Wales Health Impact Assessment Support Unit,[7] and the U.S. Health Impact Project.[8] Furthermore, the World Health Organization has compiled an Impact Assessment Directory listing references and resources.[9]

METHODS AND PROCESS OF HIA

HIAs can vary in terms of their timing and depth. They can be undertaken before implementation of a proposal (*prospectively*), during implementation (*concurrently*), and after implementation (*retrospectively*). Prospective HIAs give the greatest opportunity for influencing change, whereas concurrent and retrospective HIAs are more monitoring and evaluation exercises, respectively. The scope of an HIA will be determined by a number of factors, including the nature and complexity of the proposal being assessed, the availability of resources, the type of data that would be needed, and the decision-making time scales.

Regarding the type of data that would be needed to conduct the HIA, information can be obtained using quantitative and/or qualitative methods. Where an estimation of the size of an impact is measurable, quantitative methods are appropriate (e.g., assessing changes to pollution particulates, changes in disease incidence). Often impacts may only be assessed through qualitative means in which individuals' experiences, perspectives, and feelings are recorded and then analyzed. This methodology is useful to assess individual concerns, anxiety, and fears, for example, and the data can be quantified for use in decision making.

Although the use of qualitative data that focuses on people's everyday experiences and perceptions is often criticized as being "anecdotal" and is often contrasted unfavorably with quantitative data, a well-designed HIA could provide a framework through which different views of evidence and health can both be made explicit and scrutinized. If, for instance, a proposed shale gas development threatens to raise levels of dust in an area, it may or may not affect respiratory health or mortality, but it could impact the home environment and affect quality of life, which is viewed as being troublesome to the local population. It is likely that a combination of quantitative and qualitative evidence will give the most holistic view of impacts. Both quantitative and qualitative studies are appropriate and useful to include in an HIA.

HIAs generally take one of three forms: *desktop*, *rapid*, or *comprehensive*. A desktop HIA may take only a few hours or a day to execute, a rapid HIA may take a few days to a few months to complete, and a comprehensive HIA is more in-depth and time- and resource-intensive, often taking many months to complete. There are five stages in the HIA process (Figure 4.1).

Screening—The screening phase is designed to take a preliminary look at the potential impacts of a proposal on the local population. In this phase, potential health risks or benefits as well as specific vulnerable groups should be identified. On the basis of the information collected from the screening phase, a decision whether to undertake an HIA can be made. If an HIA is deemed feasible and necessary, the screening phase would identify the main issues to focus on, as well as what type of HIA should be undertaken.

Scoping—This is the stage when decision-making time scales and geographic boundaries are established, resources and responsibilities are clarified, the type and methods of assessment are agreed upon, and key stakeholders (and methods of involving them) are identified.

Appraisal—This is an important stage of an HIA. In the appraisal phase, information is gathered to quantify the nature, size, significance, likelihood, and distribution of health impacts. On the basis of a statistical profile, this stage also provides an opportunity for suggesting possible ways of maximizing health benefits and minimizing health risks, particularly for those cohorts deemed most vulnerable to the proposed action. During the appraisal phase, it is helpful to make use of available, appropriate, and attainable baseline information, taking into account time and resource constraints. For example, a population profile based on existing data is a useful means of providing a snapshot of the area that would be affected by the proposed action. It should be made clear that an HIA is not in itself a research methodology; rather, it draws on a range of sources of information and methods, both quantitative and qualitative, for collecting and analyzing data.

Reporting and Recommendations—Once evidence and data collection are complete, a set of clear, concise, realistic, and achievable recommendations should be developed. The findings of the HIA should be collated into a report that would be accessible to key stakeholders.

Monitoring and Evaluation—Monitoring and evaluation, integral parts of the HIA process, are unfortunately often overlooked. Because the purpose of an HIA is to inform decision making, it is important to evaluate its effectiveness in influencing decisions. After all, the primary purpose of an HIA is to present unbiased information on which recommendations would be made.

Figure 4.1
The HIA Process.

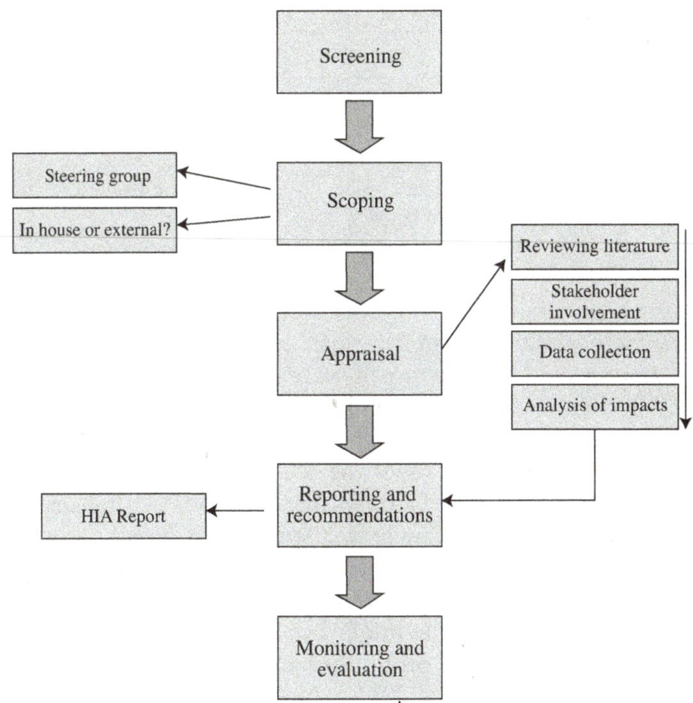

Source: Chadderton et al.[10]

TYPES OF HIAS

There are two primary types of HIAs. The "tight" HIA, which is epidemiologically focused (e.g., relies on quantifiable health impacts such as air quality and emissions, water quality, and noise level), and the "broad" HIA, which is more sociological in nature in that the focus tends to be on the wider societal determinants of health, including health inequalities and lay knowledge. This perspective is grounded in a robust mixed methodology of qualitative and quantitative evidence including local community knowledge. Table 4.1 summarizes the differences between the two perspectives.

POTENTIAL HEALTH IMPACTS OF UGE: WHAT DOES THE EVIDENCE SHOW?

There has been a considerable amount of discussion and debate over the pros and cons of UGE, and the chapters in this book highlight a substantial number of these issues. More articles on UGE have been published in the

Table 4.1
"Broad" Perspective versus "Tight" Perspective[11]

	Perspective	
Differences	**Broad**	**Tight**
Precision	Low	High
Emphasis	Sociology	Epidemiology
Ethos	Democratic	Technocratic
Quantification	Qualitative	Quantitative
Evidence	Key stakeholders	Statistics
Effects	Positive and negative	Risk focused

peer review medical literature in 2014 than in 2011 and 2012 combined.[12] It should be emphasized that there is a wide range in the quality of these studies, and findings may vary depending on the composition of the study population, the type of data collected, and so forth. Systematic reviews generally offer the best overview of the scientific research data, but care must be taken in the selection of studies for inclusion. Comparing "apples to oranges" will not advance knowledge. Appendix A lists the main potential health effects of UGE that have been identified in the literature so far.

There is a growing body of literature on the potential for harm to human health from UGE.[13–15] The Public Health implications of UGE were discussed at the Fifth Annual Summit of the Research Triangle Environmental Health Collaborative in 2012 in North Carolina.[16] The consensus at the time was that research on the potential effects of UGE on human health was in its infancy and that there was a need for more research to better understand the short- and long-term effects on human health. The conference highlighted the need for more information, including the need for comprehensive baseline health data, the identification of economic and community infrastructure impacts, and management best practice to safeguard public health.

Public Health England issued a comprehensive report in 2013 and came to the conclusion that the potential risks to public health from exposure to the emissions associated with shale gas extraction are low if the operations are properly run and regulated.[17] Some researchers have taken exception to this conclusion.[18] Another review highlighted the lack of academic research on UGE and concluded that based on the evidence it is not possible to say whether this process is or is not detrimental to health and well-being.[19] Clearly, much more information is needed to draw conclusions about the benefits and risks of UGE.

Within the context of limited epidemiologic research, the number of HIA studies conducted is equally small. As of this writing, there has been only one well-designed, comprehensive HIA published in a peer-reviewed journal.[20] This study focused on shale gas development in Colorado, and it assessed the development from a broad HIA perspective. (See Appendix B for a summary of this study.)

ADVANTAGES AND CHALLENGES ASSOCIATED WITH USING HIA IN SHALE GAS EXTRACTION

There are many advantages to requiring an HIA to assess the impacts of proposed shale gas developments, including increasing awareness across sectors of how shale gas extraction could affect health and well-being; bringing together stakeholders in an open way to discuss issues of concern; coordinating action between sectors to protect health and well-being; promoting evidence-based planning and decision making; and giving a clearer view of what is being planned and what impact there might be on the community. HIAs should be an integral component in assessing the benefits and harms of UGE. Ideally, they should be conducted before any drilling, and the findings should be used in the planning process to minimize the potential for harm. Involving communities and key stakeholders (e.g., industry, government, regulatory agencies, health professionals, and the local community) is vital, and HIA can be used as one of a range of community engagement tools.

Published epidemiologic studies focusing on health outcomes are virtually nonexistent, which makes it challenging to scientifically validate anecdotal reports of health outcomes. Relying only on anecdotal reports is not helpful unless validated by empirical studies.[21]

Although the definition of HIA is generally agreed on, complicating matters is the fact that countries have different legislative/regulatory frameworks and public health systems with different priorities. As such, HIAs have to be designed within these contexts and constraints. Furthermore, funding for HIAs has been problematic, thus creating the situation in which those with vested interests (e.g., the oil and gas industry) can and do fund the study as part of a statutory obligation to conduct an EIA. Furthermore, lack of public health resources and capabilities to undertake HIAs has driven a growth in international HIA consultants. If an HIA is commissioned, the likelihood is that it will be paid for by the developer and will be undertaken alongside the statutory obligation to conduct an EIA. Clearly this could be construed as a conflict of interest and compromise the study findings. Skepticism and mistrust, particularly with regard to planning decisions, would be difficult to overcome.

Another challenge is how to handle the local community's expectations of what an HIA can and cannot achieve. One must clearly set out what the aims and objectives of the HIA are from the outset and involve stakeholders in the early planning process and give them a voice. It would be a shame if the HIA was used as a political protest tool by community activists. The HIA should be promoted as a mechanism to collect information for use in and to support the planning processes.

CONCLUSION

The U.S. National Academy of Sciences has endorsed the concept of the HIA.[22] Public Health England also has gone on the record as endorsing such studies before shale gas developments are allowed to proceed.[23] HIAs should be viewed as a valuable instrument to support effective decision making, especially in regard to UGE. A broad, robust, and inclusive HIA should involve stakeholders in an open, democratic, and transparent way. Some authorities have taken the first steps to ensure that HIAs are conducted. For example, in New Brunswick, Canada, the Chief Medical Officer of Health has specifically recommended the use of HIA as part of the regulatory process in relation to shale gas developments.[24] The governor of New York banned UGE based on the Department of Health's assessment of the level of risk to public health and to the environment.[25] Perhaps as interest in UGE goes global, the realization that it is always better to have data before endorsing a potentially harmful course of action will (finally) gain acceptance.

NOTES

1. European Centre for Health Policy. *Health Impact Assessment: Main Concepts and Suggested Approach.* Gothenburg consensus paper. Brussels: WHO European Centre for Health Policy, 1999.
2. Scott-Samuel A, Birley M, Ardern K. *Merseyside Guidelines for Health Impact Assessment.* Liverpool, United Kingdom: International Health IMPACT Assessment Consortium, 2001.
3. Welsh Assembly Government. *Minerals Technical Advice Note 2: Coal.* 2009.
4. European Centre for Health Policy, *Health Impact Assessment,* op. cit.
5. International Association of Impact Assessment website. http://www.ihia.org.uk (accessed July 28, 2014).
6. HIA Gateway website. http://www.apho.org.uk/default.aspx?QN=P_HIA (accessed July 28, 2014).
7. Wales Health Impact Assessment Support Unit website. http://www.whiasu.wales.nhs.uk (accessed July 28, 2014).
8. Health Impact Project. http://www.healthimpactproject.org (accessed July 28, 2014).
9. World Health Organization HIA website. http://www.who.int/water_sanitation_health/resources/hia/en (accessed July 28, 2014).
10. Chadderton C, Elliott E, Williams G. Involving the public in HIA: an evaluation of current practice in Wales. *Sociology of Health and Illness* 31 (2008): 1101–1116.
11. Green L. *Health Impact Assessment of a Health Precinct.* Master's thesis, Manchester University, Manchester, United Kingdom, 2008.

12. Compendium of scientific, medical, and media findings. www.concernedhealthyny.org/compendium (accessed August 6, 2014).

13. Public Health England. *Review of the Potential Public Health Impacts of Exposures to Chemical and Radioactive Pollutants as a Result of Shale Gas Extraction*. London: Public Health England, 2013.

14. Adgate JL, Goldstein BD, MacKenzie LM. Potential public health hazards, exposures and health effects from unconventional natural gas development. *Environmental Science and Technology* 48, no 15 (2014): 8307–8320.

15. Atherton F. *Discussion Paper: Hydraulic Fracturing and Public Health in Nova Scotia*. Nova Scotia Hydraulic Fracturing Independent Review and Public Engagement Process, 2014.

16. Tillett T. Summit discusses public health implications of fracking. *Environmental Health Perspectives* 121, no. 1 (2013): a15. http://ehp.niehs.nih.gov/121-a15 (accessed August 6, 2014).

17. Public Health England. *Review of the Potential Public Health Impacts of Exposures to Chemical and Radioactive Pollutants*, op. cit.

18. Law A, Hays J, Shonkoff SB, Finkel ML. Public Health England's draft on shale gas extraction. *BMJ* 348 (2014): g2728. doi: 10.1136/bmj.g2728.

19. Adgate et al., Potential public health hazards, op. cit.

20. Witter R, McKenzie L, Towle M, et al. *Health Impact Assessment for Battlement Mesa, Garfield County Colorado*. Denver: Colorado School of Public Health. 2011. http://www.garfield-county.com/public-health/documents/1%20%20%20Complete%20HIA%without%20Appendix%20D.pdf (accessed August 1, 2014).

21. Adgate et al., Potential public health hazards, op. cit.

22. Institute of Medicine. *Health Impact Assessment of Shale Gas Extraction: Workshop Summary*. Washington, DC: The National Academies Press, 2014.

23. Public Health England. *Review of the Potential Public Health Impacts of Exposures to Chemical and Radioactive Pollutants*, op. cit.

24. Province of New Brunswick. *Chief Medical Officer of Health's Recommendations Concerning Shale Gas Development in New Brunswick*. 2012. http://leg-horizon.gnb.ca/e-repository/monographs/31000000047096/31000000047096.pdf (accessed July 28, 2014).

25. New York State Department of Health. A Public Health Review of High Volume Hydraulic Fracturing for Shale Gas Development, 2014. www.health.ny.gov.

APPENDIX A

ENVIRONMENTAL AND PHYSICAL HEALTH IMPACTS

Groundwater contamination: This is most likely the result of leakage through vertical boreholes, well casing, and water pollution; could potentially occur via migration through fractured rock, by surface spills of fracturing fluids or wastewater, or through accidents and malfunctions. Abandoned wells present another potential pathway for water pollution.[1,2]

Atmospheric pollution: This is most likely from traffic and machinery emissions and from fracturing contaminants combining with other pollutants. Airborne chemicals can have a negative impact on health through either direct contact or via inhalation. For example, ozone is recognized as a contributing factor for respiratory and cardiovascular health-related issues whereas silica, a proppant, can turn to dust and is carcinogenic if inhaled. Condensate tanks emit noxious compounds, including carbon disulfide, which over time can cause cardiovascular, neurologic, and hepatic effects.[3]

HEALTH, WELL-BEING, AND QUALITY-OF-LIFE IMPACTS

Adverse birth outcomes: There is evidence of an association between maternal residential proximity to natural gas developments and congenital heart defects and possibly neural tube defects.[4]

Noise: The noise from drilling, compressor stations, flaring, and diesel generators as well as heavy truck traffic can contribute to an increase in stress and raise blood pressure.[5]

Psychological impacts: Mental health issues include an increase in depression, anxiety, and stress.[6]

Traffic accidents: Increase in traffic on rural roads increases the likelihood of road accidents and fatalities.[7]

Social and community impact: Factors include a disruption in the rural fabric of the community as well as the increase of individuals drawn to the community for employment. This dynamic changes the demographics of the locale.[8]

Economic impacts: Well-paid workers and the need for housing changes the dynamics of the community, with many locals being priced out of living in town.[9]

NOTES

1. Public Health England. *Review of the Potential Public Health Impacts of Exposures to Chemical and Radioactive Pollutants as a Result of Shale Gas Extraction.* London: Public Health England, 2013.

2. Adgate JL, Goldstein BD, MacKenzie LM. Potential public health hazards, exposures and health effects from unconventional natural gas development. *Environmental Science and Technology* 48, no. 15 (2014): 8307–8320.

3. Schmidt CW. Blind rush? Shale gas boom proceeds amid human health questions. *Environmental Health Perspectives* 119, no. 8 (2011): A348–A353.

4. McKenzie LM, Guo R, Witter R, et al. Maternal residential proximity to natural gas development and adverse birth outcomes in rural Colorado. *Environmental Health Perspectives* 122, no. 4 (2014). http://ehp.niehs.nih.gov/1306722 (accessed November 8, 2014). doi:10.1289/ehp.1306722.

5. Schmidt, Blind rush?, op. cit.

6. Witter R, McKenzie L, Towle M, et al. *Health Impact Assessment for Battlement Mesa, Garfield County Colorado*. Denver: Colorado School of Public Health. 2011. http://www.garfield-county.com/public-health/documents/1%20%20%20Complete%20HIA%without%20Appendix%20D.pdf (accessed August 1, 2014).

7. Tillett T. Summit discusses public health implications of fracking. *Environmental Health Perspectives* 121, no. 1 (2013): a15. http://ehp.niehs.nih.gov/121-a15 (accessed August 6, 2014).

8. Witter et al. *Health Impact Assessment for Battlement Mesa*, op. cit.

9. Atherton F. *Discussion Paper: Hydraulic Fracturing and Public Health in Nova Scotia*. Nova Scotia Hydraulic Fracturing Independent Review and Public Engagement Process, 2014.

APPENDIX B

HIA FOR BATTLEMENT MESA, GARFIELD COUNTY, COLORADO, 2010

This prospective, comprehensive HIA, initiated at the request of the Garfield County Board of County Commissioners in response to a citizen petition, was a collaborative study undertaken by the Department of Environmental and Occupational Health, Colorado School of Public Health and Garfield County Public Health.[1] The purpose of the HIA was to assess the potential impacts of Antero Resources Corporation's proposal to construct 200 natural gas wells and associated facilities/infrastructure in the residential community of Battlement Mesa and to provide health information and recommendations to the BOCC. The geographic distribution of the proposed development indicated that effects could be community-wide.

The HIA used both quantitative and qualitative sources of information. Following screening and scoping, the HIA focused on eight specific areas of health concern (stressors): air emissions; water and soil quality; transportation and traffic; noise, vibration, and light pollution; community wellness; economics and employment; health infrastructure; and accidents and malfunctions.

The assessment included construction of a baseline demographic and health profile for Battlement Mesa, a health literature review, and a human health risk assessment using longitudinal air and water quality data. Each stressor was assessed in terms of current evidence, current conditions, changes likely to occur as a result of the proposed drilling plans, and potential health impacts. Each stressor assessment concluded with a summary of findings and list of recommendations. Recommendations included the establishment of a community advisory board independent of the industry, building new roads to separate industrial traffic from community, and to develop robust emergency response plans, continued monitoring of air emissions, and funding baseline/ongoing monitoring of certain community wellness measures (e.g., land values, substance abuse, sexually transmitted diseases, school climate and enrollment, psychosocial status of community members) to ensure implementation of actions to address adverse impacts. The HIA report outlined next steps and conclusions and included a number of supporting documents as appendices.

In 2013, a reflection of the experience of undertaking the Battlement Mesa HIA highlighted the gaps, challenges, lessons learned, and utility of the process.[2] Challenges related to the availability of and ongoing additions and modifications to operator plans and information, stakeholder selection, funding, and political pressures. Lessons learned highlighted the requirement for the careful vetting of stakeholders, adherence to mutually agreed deadlines, agreeing on procedures for evaluating new information, and equality of access to stakeholder meetings. Encouraging the independent funding of such HIAs and understanding the context in which they are undertaken were also cited. For financial and political reasons, the Battlement Mesa HIA was not finalized, but it serves as a useful guide for evaluating the health impacts of natural gas–related development within residential communities.

Overall, the main findings of the paper suggest that more extensive research on environmental emissions, community impacts, and health outcomes should be conducted to allow for a full understanding of health impacts relating to natural gas development. Temporary moratoriums have been imposed by some local and state community governments in response to an absence of data on UGE impact. In such circumstances and where scientific information is incomplete, HIA methods provide guidance to community leaders regarding the inclusion of health in the decision-making process. There must be an open and transparent information-gathering and analysis process that involves local stakeholders.

ACKNOWLEDGMENT

I acknowledge and thank my colleague Julia Lester at the Public Health Wales for her ideas and suggestions.

NOTES

1. Witter R, McKenzie L, Towle M, et al. *Health Impact Assessment for Battlement Mesa, Garfield County Colorado.* Denver: Colorado School of Public Health. 2011. http://www.garfield-county.com/public-health/documents/1%20%20%20Complete%20HIA%without%20Appendix%20D.pdf (accessed August 1, 2014).

2. Witter RZ, McKenzie L, Stinson KE, et al. The use of health impact assessment for a community undergoing natural gas development. *American Journal of Public Health* 103, no. 6 (2013): 1002–1010. doi:10.2105/AJPH.2012.301017.

Chapter 5

Natural Gas Development and Its Effect on Air Quality

Nathan P. De Jong, Roxana Z. Witter, and John L. Adgate

INTRODUCTION

Large-scale unconventional petroleum development and production has expanded rapidly since 2007 in the United States, resulting in the drilling of new wells in established production regions as well as in new areas. As a result, it is estimated that more than 15 million Americans live within a mile of a well drilled since 2000.[1] The resulting production wells, equipment, and accompanying infrastructure are sources of a wide range of air pollutants that have potential impacts on public health at the local and regional scale. The wells themselves may emit methane, aromatic and aliphatic hydrocarbons, and, depending on the formation, hydrogen sulfide (H_2S). Equipment needed to construct well infrastructure, fracture the underlying formations, capture waste, and prepare the petroleum products for distribution are all potential sources of emissions of particulate matter (PM), silica, diesel exhaust, nuisance dusts, and odors.[2–4] Diesel engines and other associated processes and equipment emit nitrogen dioxides (NOx), polycyclic aromatic hydrocarbons (PAHs), particulate matter, and other pollutants. All these sources may also contribute to or exacerbate regional ground-level ozone formation.

This chapter explores the overall impact of unconventional oil and gas extraction on air quality, with a focus on the impact of the recent boom in natural gas development. We refer to the process as unconventional gas extraction (UGE), as is done in the other chapters in this book. To truly understand the impact UGE has on air quality, one needs to understand the phases of the production process. We present the six main categories of air pollutants and the health hazards associated with them. We also review the current, limited information about human exposure and epidemiological data related to these exposures. Lastly, we describe the knowledge gaps and research opportunities.

AIR EMISSIONS FROM THE UNCONVENTIONAL PETROLEUM DEVELOPMENT AND PRODUCTION PROCESS

Preproduction Phase

UGE is an industrial process best viewed as a series of steps with the potential for air emissions at numerous points throughout the process. The initial phase of development is termed *preproduction*.[5] Preproduction includes permitting under the relevant state and/or local jurisdictions, petroleum exploration, well pad selection, site preparation, drilling, hydraulic fracturing, and well completion. For a single well, preproduction is usually completed within a few months. When these operations are carried out for multiple wells on a pad and at multiple sites preproduction can last several months to more than 1 year.[6]

In the first step of preproduction—petroleum exploration—potential drilling sites are identified through geologic studies, seismic interpretation, and a petrophysical assessment.[7] Once a site for a well pad is chosen, site preparation and large industrial activity begins with hauling in materials to construct the drill rig and install associated infrastructure. Site preparation includes the construction of roads and holding ponds, the clearing and leveling of the well pad, and the placement of pipelines for natural gas transport during production. During site preparation, large diesel-powered trucks bring materials to and from the development site.[8,9] Diesel-powered trucks emit diesel exhaust containing ozone precursors NOx and volatile organic compounds (VOCs), as well as fine particulate matter ($PM_{2.5}$) and PAHs.[10,11] On- and off-road truck traffic as well as tire and brake wear generate coarse particulate matter (i.e., <10 micrometers in diameter [PM_{10}]).[12]

Diesel-powered industrial equipment is used to construct the infrastructure of the well and prepare it for the next phase of preproduction development, vertical and horizontal drilling into the production zone of the unconventional resource "play." Drilling uses heavy equipment to insert the drill bit, pipe, and casing into the reservoir where the subsurface hydrocarbons are contained. There are no published studies on the air hazards from drilling fluids and muds, although some authors have noted the potential for occupational exposures from vaporization or aerosolization of drilling fluids.[13,14]

Once drilling is completed, the next stage of preproduction, well completion, can begin. Well completion activities are interlinked with the drilling of the wellbore; data are gathered on the formation that will be used to design the well completion process and finalizing well casing, which is set and cemented. The blowout preventer is replaced with wellhead control valves, and connections are made to the production facilities. Hydraulic fracturing then uses heavy equipment to pump water, fracturing fluids, and proppant (usually silica-based sand) underground at high pressures to create pathways for the trapped hydrocarbons to come to the surface. Millions of gallons of water, sand, and hydraulic fracturing chemicals are transported to and from the well pad site, and trucks and equipment contribute a wide range of pollutants to local and regional air sheds. During this phase, both the well and associated equipment used in the drilling may be major sources of air pollution, depending on the emissions control technology used.

After a well has been fractured, hydraulic fracturing fluids "flowback" to the surface. Flowback water, primarily composed of the water and chemicals used in hydraulic fracturing, contains subsurface liquids and solids, including liquid hydrocarbons (condensate) and natural gas. The composition of flowback water depends on local basin characteristics, hydraulic fracturing fluid components, the content of the subsurface water contained in the play, and oil and natural gas contained in the reservoir.[15] Both the well itself and flowback fluids may be a significant source of fugitive methane and a range of hydrocarbons, engine-related, and other hazardous air pollutants (HAPs).[16]

Venting occurs during well completions, hydraulic fracturing, and production. With venting, there is an intentional release of methane and other nonmethane hydrocarbons into the atmosphere through equipment design or operator practice. Flaring, the controlled burning of natural gas, also occurs before the well is connected to a transmission pipeline. Flaring of the natural gas mixture is used so that methane and other nonmethane hydrocarbons are not released directly into the air shed; however, combustion emissions are released, including NOx, sulfur oxides (SOx), formaldehyde, noncombusted methane, VOCs, and HAPs.[17] The quantity and composition of emissions at this stage is highly variable depending on the methods and care used by the companies. Venting and flaring of natural gas is a source of emissions during the well completion process, which includes benzene, toluene, ethylbenzene, and xylenes (BTEX) and other HAPs; methane and other nonmethane hydrocarbons; and H_2S.[18]

Once the natural gas becomes saleable, connections to the transportation line are made, and the well is transformed into a producing well.

Production Phase

Production operations encompass all actions taken to achieve the maximum recovery of hydrocarbons from the well (both gas and liquid) and a removal of produced water. Surface gases are separated from liquids; the liquids are collected and separated into produced water and liquid hydrocarbons (oil and/or condensate) and stored on site until transport is available. Natural gas is compressed on site and/or sent to centralized gas compressors. These compressors are an additional source of leaked natural gas mixture into the air.[19,20]

There are a number of emission point sources at the well pad, for example, well-head pumps that bring the produced gas to the surface, pipeline joints and valves, dehydrator regeneration vents, pneumatic pumps, and compressor stations.[21] Some emissions may be captured and flared, resulting in combustion emissions and incomplete capture or burning in flaring systems that may contribute to HAPs and methane emissions. Trucks used onsite during production are also emission sources that contribute VOC, NOx, and particulate matter. Fugitive emissions can also come from well pad equipment bleeding and leaks as well as leaks from well casings. Product handling represents another potential source of large, intermittent emissions of methane, aliphatic, and aromatic hydrocarbons. These emissions can be continuous or intermittent but will be ongoing during the entire lifetime of the well. Air emissions from a large number of wells in a play may increase the negative impact on local and regional air quality.

Transmission, Storage, and Distribution

Much has been said about the "methane issue" associated with natural gas extraction. It is important to understand that natural gas is predominantly methane, a powerful greenhouse gas with important implications for greenhouse gas reduction and climate change.[22] Evidence indicates that methane leaks can occur throughout the natural gas transmission system contributing to the overall impact of the UGE on global climate.[23,24]

Well Maintenance

Well maintenance, or well workovers, can lead to increased, episodic emissions from existing wells. Well workovers are the process of performing major maintenance or remedial treatments on an oil or gas well. It includes the repair or refracturing of an existing producing well to restore, prolong, or enhance the hydrocarbon production. The process of "pigging" involves opening gas lines for cleaning, sometimes without stopping the flow of gas in the pipeline. All of these operations can be significant sources of episodic VOCs, HAPS, methane, and other petroleum hydrocarbons.

Emission Control Technology

The U.S. Environmental Protection Agency (EPA) recently issued oil and natural gas air pollution standards for reduced emission or "green" completions, which are designed to reduce 95% of VOC emissions.[25] Current practice involves capturing and selling the gas once it is considered salable, but early capture during flowback requires investment in gas separation and processing facilities, which at this time may not be considered technically or economically feasible.[26] That being said, the EPA has issued a series of reduced emissions completions for hydraulically fractured natural gas wells that are to go into effect in 2015.[27,28] Compliance with these rules will likely reduce occupational and community exposures to VOCs from this source. It may also have a positive regional effect on air quality by reducing emissions of smog forming VOCs.

Table 5.1 lists air pollution sources, activities, and impacts associated with UGE.

MAJOR CLASSES OF POLLUTANTS, HUMAN EXPOSURES, AND ASSOCIATED HEALTH EFFECTS IN WORKERS AND COMMUNITIES

The process of UGE releases a wide range of chemical and physical hazards that affect air quality on a local, regional, and global scale. In this section, we describe the current state of the science on emissions and health effects for the main air pollutant hazards that stem from UGE: (1) petroleum hydrocarbons, (2) silica, (3) diesel particulate matter, (4) radiation, (5) hydrogen sulfide gas, and (6) ozone.

Petroleum Hydrocarbons

A wide range of methane and nonmethane hydrocarbons have been measured in and around well sites, which can have effects on regional air quality, particularly around large production fields.[29–34] Studies in Colorado's Piceance Basin, Pennsylvania's Marcellus Shale, and Texas's Barnett Shale have linked well completions and associated infrastructure such as compressors, condensate storage tanks and dehydrators, and other infrastructure to increases in atmospheric VOCs and other HAPs.[35–38] Measurements during uncontrolled flowback in Garfield County, CO, for example, found that VOCs were detected more often and at higher concentrations than in regional air quality measurements.[39] Another peer-reviewed study exploring

Table 5.1
Air Pollution Sources, Activities, and Impacts Associated with Unconventional Gas Extraction

Source	Activities	Air Impact
Large trucks	All	Diesel emissions
Industrial equipment	All	Diesel emissions
Drilling muds and cuttings	Well construction; drilling	Particulate matter; volatile drilling muds; petroleum hydrocarbons
Fracturing fluid	Hydraulic fracturing; flowback	Silica; volatile fracturing fluids
Generators and pumps	Drilling; hydraulic fracturing; flowback	Diesel emissions
Flowback water	Flowback	Fracturing fluids volatile; petroleum hydrocarbons
Produced water	Well drilling; flowback	Drilling muds volatile; petroleum hydrocarbons
Gas venting	Well drilling; hydraulic fracturing; flowback; production; well maintenance	Methane and other nonmethane hydrocarbons
Gas flaring	Well drilling; flowback; production	NOx, CO_2
Petroleum production	Production	Petroleum hydrocarbons
Condensate tanks	Production	Methane; petroleum hydrocarbons
Well workovers and pigging	Well maintenance	Methane; petroleum Hydrocarbons
Pipelines	Production, TSDU	Methane; petroleum hydrocarbons

TSDU = transmission, storage, distribution, use.

emissions at a site using "green" (or reduced emission) completion found that 24-hour integrated air samples collected 0.7 miles from a well pad measured higher average emissions during drilling than during the green completion, demonstrating that hydrocarbon emissions may also need to be controlled at this stage in the development process.[40] Additional studies on the effect of control technologies are needed to understand more comprehensively the impacts of "green" completion on reducing hydrocarbon emissions and human exposure.

Regarding the link between exposures and health effects, workers may be exposed to a mixture of petroleum hydrocarbons during the many activities of preproduction and production phases of extraction. A report prepared for the Texas Commission on Environmental Quality by Hendler et al. documented speciated measurements of VOCs from 11 oil and 22 condensate storage tanks at the wellhead and gathering site tank batteries in the Houston-Galveston-Brazoria (HGB), Dallas-Fort Worth (DFW), and Beaumont-Port Arthur (BPA) of the Eagle Ford Shale area of East Texas during May–July 2006. The researchers were sampling vented gas emissions from the onsite equipment in the area where employees could be exposed to episodic hydrocarbons. The arithmetic mean of the total uncontrolled VOC emission estimates for wellhead and gathering site storage tanks in HGB, DFW, and BPA were 289 tons, 38 tons, and 145 tons per day, respectively.[41]

Workers in close proximity to flowback operations may have more significant occupational exposure to VOCs and HAPs, particularly during venting and flaring before the well is connected to a transmission pipeline. Esswein et al. performed direct sampling during flowback operations at six sites in Colorado and Wyoming during the spring and summer of 2013 and identified benzene as the primary VOC exposure hazard for workers.[42] Fifteen of 17 personal breathing zone (PBZ) samples met or exceeded the National Institute for Occupational Safety and Health (NIOSH) Recommended Exposure Limit (REL) of 0.1 ppm as a full shift Time Weighted Average (TWA) for workers gauging flowback or production tanks. The measurements of airborne VOC exposure for workers on well sites and near an assortment of point sources approached concentrations that potentially pose health risks for workers. On the basis of these initial findings, the authors determined that opening thief hatches and gauging tanks is the primary task-based activity that increases inhalation exposure risks, as time spent working around flowback and production tanks increases.

There is evidence of an increased risk of multiple myeloma and non-Hodgkin's lymphoma and noncancer health effects including asthma exacerbation, headaches, and mucous membrane irritation.[43–45] Prolonged benzene exposure, for example, is associated with acute and chronic nonlymphocytic leukemia, acute myeloid leukemia, chronic lymphocytic leukemia, non-Hodgkin's lymphoma, anemia, and other blood disorders and immunological effects.[46] Toxicity information for other hydrocarbons, such as heptane, octane, and diethylbenzene, is limited, hindering the risk assessment for exposures to these compounds.[47]

There are documented instances of odor complaints during various phases of UGE development. Eighty-one percent of respondents to a self-reporting survey in active shale gas development areas in Pennsylvania's Marcellus Shale play reported odors form UGE.[48] Hydrogen sulfide gas (H_2S) may be responsible for some odor complaints.[49] People living near natural gas operations also have reported upper respiratory, neurological, and dermatological symptoms that appeared after drilling.[50–52] Although these studies have limitations because they are convenience samples of the local population, these effects are consistent with known health effects associated with petroleum hydrocarbon exposure. For example, inhalation of trimethylbenzenes and xylenes can irritate the respiratory system with effects ranging from eye, nose, and throat irritation to difficulty in breathing and impaired lung function.[53,54] Inhalation of xylenes, benzene, and aliphatic hydrocarbons can adversely affect the nervous system with effects ranging from dizziness, headaches, fatigue, and numbness in the limbs to lack of muscle coordination, tremors, temporary limb paralysis, and unconsciousness at high levels.[55]

Few studies have been conducted to explore the association between development and birth outcomes and childhood cancer. A retrospective cohort study of 124,862 live births between 1996 and 2009 in rural Colorado reported an association between maternal proximity to natural gas wells and the birth prevalence of congenital heart defects and neural tube defects, but no association with oral clefts, low term birth weight, or preterm birth weight.[56] Again, there are limitations to this study including potential confounding factors, but the findings are troubling enough to warrant further investigation. Conversely, Fryzek et al. compared standardized incidence rates (SIRs) for childhood cancer in Pennsylvania counties and reported no difference between SIRs for all cancers except central nervous system tumors in the counties with the fewest wells.[57] Limitations of this study include insufficient consideration given to latency period needed for cancer development, lack of an individual level assessment of relevant confounders, and assumptions made about individual exposures. Overall, these early epidemiological studies lack spatial and temporal specificity in individual-level risk factors and exposure. They underscore the need for more rigorous peer-reviewed studies examining health effects in populations living in and around UGE areas.[58]

Silica

Silica sand, the most common additive in fracturing fluid, is used as a proppant to hold open (prop) fractures, allowing gas to escape from the gas-bearing formations. Mechanical handling of fracking sand creates large clouds of respirable silica dust.[59] One of the key occupational hazards for workers on well pads is emission of respirable crystalline silica released during the mechanical handing of fracturing sand. Esswein et al. studied workers in Colorado, Texas, North Dakota, Arkansas, and Pennsylvania and found that 12-hour time weighted average breathing zone silica samples exceeded the American Conference of Industrial Hygienists threshold limit value of 0.025 mg/m^3, and most (68.5%) exceeded the NIOSH REL of 0.05 mg/m^3. Fifty-seven of the 111 (51.4%) samples exceeded the Occupational and Safety Health Administration's (OSHA) permissible exposure limit (PEL) for respirable silica containing dust. Workers who inhale respirable silica are at an increased risk for developing other silica-related diseases, such as chronic obstructive pulmonary disorder (COPD), kidney disease, and autoimmune disease. Silica also increases the risk of tuberculosis.[60]

There is strong evidence that current OSHA silica limits are not adequately protective of worker health. The current OSHA PEL is based on research from the 1960s and does not reflect more recent scientific evidence, which has led OSHA to recently propose lowering its PEL to match the NIOSH REL of 0.050 mg/m^3.[61] OSHA estimates that if the proposed rule takes full effect, approximately 700 lives could be saved and 1,600 new cases of silicosis prevented per year.

Dust from construction activities, well pads, and access roads also has the potential to affect local air quality. That is, coarse particulate matter (PM_{10}) can be emitted from dust suspended by trucks traveling on unpaved roads as well as from dust generated by tire wear and brake wear.[62] Although PM_{10} is not likely to be a health concern primarily because PM_{10} cannot reach the gas exchange regions of the lung, it can be a substantial nuisance and have a negative impact on quality of life for those living near development areas.

Diesel Particulate Matter

Diesel exhaust is a complex mixture of gases and particulate combustion products that is affected by the nature of the engine, operating conditions, lubricating oil, additives, emission control systems, and fuel composition (e.g., heavy-duty diesel engines emit more particulate matter than light-duty diesel engines). The combustion products of diesel engines are primarily composed of nitrogen, oxygen, carbon dioxide, and water vapor as well as pollutants such as carbon monoxide (CO), SOx, NOx, PM, VOCs, and low-molecular-weight PAHs.[63] Furthermore, a major component of diesel particulate matter emitted by diesel-powered engines is black carbon, which may be an important driver for climate change.[64]

In the preproduction phase of development, drill rigs and hydraulic fracturing pumps are powered by off-road heavy-duty diesel engines that emit NOx, $PM_{2.5}$, and VOCs. A single drill rig has five to seven independent diesel-powered compression ignition engines, and these engines are major sources of NOx, $PM_{2.5}$, and sulfur dioxide (SO_2).[65] Typically, there are eight to ten fracturing pumps per well powered by diesel engines, so this is not an insignificant source of pollution. Additionally, it is estimated that the number of truck trips necessary during the phases of UGE ranges from several hundred to more than a thousand per well. The EPA has classified diesel exhaust as likely to be carcinogenic in humans, and the International Agency for Research on Cancer has classified diesel exhaust as a human carcinogen.[66–68]

Currently, there are no direct measurements of occupational diesel exposure among UGE workers; however, evidence shows that exposure to diesel exhaust is associated with respiratory and cardiovascular disease in other industries.[69–71] Among those living in close proximity to drilling operations, concentrations of $PM_{2.5}$, NOx, PAHs, and VOCs from diesel at homes near well pads and truck routes have not been measured, so the magnitude of exposure from this source is not known.

Radiation

Naturally occurring radioactive materials (NORMs) are naturally occurring radionuclides that are found within the geologic formations that contain oil and natural gas deposits; examples include uranium and its decay products (including radon), thorium and its decay products, radium and its decay products, and lead-210. Natural gas deposits tend to have higher concentrations of radioisotopes, and geologists have used this as a proxy for identifying natural gas deposits.[72] Oil and gas extraction practices bring NORMs to the surface in produced water, which is created when oil and gas are separated from water and collected in tanks or pits. As oil and gas is removed from the formation, these fields generate more produced water. The extraction process concentrates and exposes the NORM to the surface environment and human contact.

The UGE extraction process also produces wastes that are classified as technologically enhanced naturally occurring radioactive material (TENORM). TENORM may pose a high risk to employees because of high concentrations of radioactive materials. Scale accumulates in drilling and production equipment as radioactive salts precipitate out of the water onto nearby solid surfaces, including well heads, casing, water lines associated with separators, heaters, treaters, and gas dehydrators.[73] Sludge from the drilling process may also contain radioactive materials. Sludge is composed of dissolved solids that precipitate out of produced water in waste pits and water storage tanks. Sludge generally consists of oil and loose material that often contains silica compounds but may also contain large amount of barium.[74]

Given the potential for harm, characterization of waste generated by UGE activities is needed and warranted to better understand risks to those working in the industry and those living in proximity to the drilling activity.

Hydrogen Sulfide Gas

"Sour gas" refers to a natural gas deposit that contains a significant amount of H_2S, which is a colorless, flammable, and extremely hazardous gas with a rotten egg smell.[75] It occurs naturally in crude petroleum and natural gas, although the frequency of its presence varies among plays. It is also produced from the bacterial breakdown of organic materials in the well bore that can develop during preproduction operations. H_2S exposures can occur during well servicing activities, tank gauging, and swabbing operations. The magnitude of occupational low level H_2S exposure has not been measured directly, but the oil and gas industry requires their employees to wear personal H_2S exposure monitors that sound an alarm if H_2S levels become immediately dangerous to life. Workers are also required to participate in H_2S training programs. OSHA recommends that natural gas employers install ground-level tank gauges and continuous monitoring during servicing operations.[76] H_2S is an explosion risk; if ignited, it produces toxic vapors and gases (e.g., SO_2).

Health effects of low-level, intermittent exposure may include irritation to the eyes, nose, or throat and difficulty breathing, especially for people with asthma. H_2S is a significant irritant, and central nervous system health effects occur at or above 100 ppm; these effects increase in severity with duration or level of exposure. Immediate death can occur at levels of ~1000 ppm.[77] Concentrations greater than 500 ppm can cause a loss of consciousness, with most persons recovering with no lasting effects. That being said, health effects of low-level, longer-term exposures are not well understood. Few data exist on the frequency, duration, and levels of occupational exposure to H_2S in the natural gas field.

Ozone

Tropospheric (ground-level) ozone is not emitted directly into the air but formed through the reaction of VOCs with NOx in the presence of sunlight.[78] Ozone precursor chemicals emitted from up and down procedures (UPDPs) operations including methane, ethane, propane, and numerous other VOCs that undergo complex photochemical reactions with NOx emissions from preproduction and production activities and other sources to ultimately produce ground-level ozone.[79] Shale gas development and production emit VOCs and NOx that can affect regional ozone levels and air quality. Studies in Texas, Colorado, and Utah have attributed emissions of light alkanes from oil and gas development to the formation and transport of ozone to nearby urban areas.[80–84]

Ozone exposure is linked to several adverse health effects, including respiratory, cardiovascular, and total mortality, as well as decreased lung function, asthma exacerbation, COPD exacerbation, cardiovascular effects, and adverse birth outcomes. Sensitive populations (e.g., those with asthma and COPD, children, and the elderly) are at an increased risk of adverse health effects.[85] People doing heavy work outside, including oil and gas workers, are likely to be more exposed to ozone than the general public and also constitute an at-risk population.

Although there are many studies documenting the health effects of ozone exposures and several studies that suggest an association between unconventional oil and gas development and ground-level ozone production, as of this writing there is only one population-based study on ozone and health effects. The study reports that from 2008 to 2011 clinics in Sublette County, Wyoming, observed a 3% increase in the number of visits for adverse respiratory-related health effects for every 10 ppb increase in the 8-hour ozone concentration from the previous day.[86] Although this increase was not statistically significant, largely due to limited statistical power, it highlights the need for further research.

The recent Allen et al. study directly measuring methane releases during the preproduction and production stages of cooperating industries in different areas of the United States is also pertinent to regional ozone production.[87] The study observed a wide range of total methane emissions within and across drilling sites.

SUMMARY AND KNOWLEDGE GAPS

Processes associated with UGE and transportation are related to air emissions that can compromise the health of workers at the well pad and community members in gas-producing areas. Workers are exposed to air pollutants on the well pad, including diesel exhaust, VOC emissions, criteria pollutants and ozone, at levels above those experienced by community members. The published literature clearly shows that there is an elevated risk to health from benzene exposure among those working in the oil and gas industry and for those living near active wells. There is an increased risk of developing chronic silicosis from exposure to silica dust. Ozone exposure has been linked with several adverse health effects, including respiratory, cardiovascular, and total mortality and decreased lung function, asthma exacerbation, COPD exacerbation, cardiovascular effects, and adverse birth outcomes

Community hazards associated with air pollutants include diesel exhaust, especially $PM_{2.5}$, from heavy trucks and diesel-powered generators; VOC emissions (particularly BTEX from flowback operations; waste and liquid hydrocarbon storage tanks, compressor, and purification equipment; leaks and releases from transportation pipelines; criteria pollutants, NOx, and SOx from gas flares and diesel combustion; and nuisance dust (PM_{10}) from traffic on unpaved roads. Emissions of ozone precursors (VOC and NOx) may lead to increased ozone formation in regions where there is heavy UGE activity. The health effects associated with these hazards range from cardiac and pulmonary effects ($PM_{2.5}$, ozone), adverse birth outcomes (BTEX, ozone), short-term irritant and neurological effects (BTEX, ozone), and the psychological effects of uncertainty about UGE exposures. Density of wells; proximity to homes, schools, and other community institutions; and cumulative emissions in regional air sheds are contributing factors to community health risks. Susceptible populations include children and fetuses, the elderly, those with chronic illnesses, and outdoor workers.

Clearly, there are potentially serious risks to health from the UGE process, and much more research needs to be done to fully understand the short- and long-term consequences of exposure (direct or indirect) to UGE activities. Improved exposure assessments and well-designed health studies will help frame the discussion about how to mitigate the potential for harm to human health and to the environment and climate.

ACKNOWLEDGMENTS

This work was supported by the AirWaterGas Sustainability Research Network funded in part by the National Science Foundation under Grant No. CBET-1240584, by the Research Partnership to Secure Energy for America/Department of Energy (RPSEA/DOE) under grant EFDTIP2-TIP213, and the Colorado School of Public Health (CSPH). Any opinion, findings, conclusions, or recommendations expressed in this chapter are those of the authors and do not necessarily reflect the views of the National Science Foundation, RPSEA/DOE, or CSPH.

NOTES

1. Gold R, McGinty T. Energy boom puts wells in America's backyards. *Wall Street Journal*, October 25, 2013. http://online.wsj.com/news/articles/SB10001424052702303672404579149432365326304 (accessed August 1, 2014).
2. Adgate JL, Goldstein BD, McKenzie LM. Potential public health hazards, exposures and health effects from unconventional natural gas development. *Environmental Science and Technology* 48, no. 15 (2014): 8307–8320. doi 10.1021/es404621d.
3. Field RA, Soltis J, Murphy S. Air quality concerns of unconventional oil and natural gas production. *Environmental Science: Processes & Impacts* 16 (2014): 954–969. doi 10.1039/c4em00081a.
4. Allen DT, Torres VM, Thomas J, et al. Measurements of methane emissions at natural gas production sites in the United States. *Proceedings of the National Academy of Sciences* 110, no. 44 (2013): 17768–17773. doi 10.1073/pnas.1304880110.
5. Branosky E, Stevens A, Forbes S. *Defining the Shale Gas Life Cycle: A Framework for Identifying and Mitigating Environmental Impacts*. Washington, DC: World Resources Institute, 2012.
6. Encana. Well Completion & Hydraulic Fracturing. http://www.encana.com/pdf/communities/usa/wellcompletionandhydraulicfracturing(DJ).pdf (accessed August 1, 2014).
7. King GE. *Hydraulic Fracturing 101: What Every Representative, Environmentalist, Regulator, Reporter, Investor, University Researcher, Neighbor and Engineer Should Know About Estimating Frac Risk and Improving Frac Performance in Unconventional*

Gas and Oil Wells. Paper presented at the SPE Hydraulic Fracturing Technology Conference, Woodlands, TX, 2012. http://fracfocus.org/sites/default/files/publications/hydraulic_fracturing_101.pdf (accessed November 9, 2014). doi 10.2118/152596-MS.

8. New York State Department of Environmental Conservation. Revised Draft SGEIS on the Oil, Gas and Solution Mining Regulatory Program: Well Permit Issuance for Horizontal Drilling and High Volume Hydraulic Fracturing in the Marcellus Shale and Other Low Permeability Gas Reservoirs; New York State, 2011. http://www.dec.ny.gov/energy/75370.html (accessed August 1, 2014).

9. *Oil and Gas: Information on Shale Resources, Development, and Environmental and Public Health Risks* (GAO-12-735). Washington, DC: U.S. Government Accountability Office, 2012. http://www.gao.gov/assets/650/647791.pdf (accessed August 1, 2014).

10. National Toxicology Program. *Diesel Exhaust Particulates*. Washington, DC: Department of Health and Human Services, 2011. http://ntp.niehs.nih.gov/ntp/roc/twelfth/profiles/DieselExhaustParticulates.pdf (accessed August 1, 2014).

11. Partial List of Chemicals Associated with Diesel Exhaust. U.S. Occupational Health & Safety Administration, U.S. Department of Labor. https://www.osha.gov/SLTC/dieselexhaust/chemical.html (accessed August 1, 2014).

12. Moore CW, Zielinska B, Petron G, et al. Air impacts of increased natural gas acquisition, processing, and use: a critical review. *Environmental Science & Technology* 8, no. 15 (2014): 8349–8359. doi 10.1021/es4053472.

13. Searl A, Galea K. *Toxicological Review of the Possible Effects Associated with Inhalation and Dermal Exposure to Drilling Fluids Production Streams*. Washington, DC: Institute of Medicine, 2011. http://www.iom-world.org/pubs/IOM_TM1104.pdf (accessed August 1, 2014).

14. Broni-Bediako E, Amorin, R. Effects of drilling fluid exposure to oil and gas workers presented with major areas of exposure and exposure indicators. *Research Journal of Applied Sciences, Engineering and Technology* 2, no. 8 (2010): 710–719.

15. Field et al., Air quality concerns, op. cit.

16. McKenzie LM, Witter RZ, Newman LS, et al. Human health risk assessment of air emissions from development of unconventional natural gas resources. *Science of the Total Environment* 424 (2012): 79–87. doi 10.1016/j.scitotenv.2012.02.018.

17. Bar-Ilan A, Grant J, Parikh R, et al. A comprehensive emissions inventory of upstream oil and gas activities in the Rocky Mountain States. 2008. http://www.epa.gov/ttnchie1/conference/ei19/session8/barilan.pdf (accessed August 1, 2014).

18. Olaguer EP. The potential near-source ozone impacts of upstream oil and gas industry emissions. *Journal of the Air & Waste Management Association* 62, no. 8 (2012): 966–977. doi 10.1080/10962247.2012.688923.

19. Rich A, Grover JP, Sattler ML. An exploratory study of air emissions associated with shale gas development and production in the Barnett Shale. *Journal of the Air & Waste Management Association* 64, no. 1 (2013): 61–72. doi 10.1080/10962247.2013.832713.

20. Roy AA, Adams PJ, Robinson AL. Air pollutant emissions from the development, production, and processing of Marcellus Shale natural gas. *Journal of the Air & Waste Management Association* 64, no. 1 (2013): 19–37. doi 10.1080/10962247.2013.826151.

21. Esswein EJ, Snawder J, King B, et al. Evaluation of some potential chemical exposure risks during flowback operations in unconventional oil and gas extraction: preliminary results. *Journal of Occupational and Environmental Hygiene* 11, no. 10 (2014): D174–D184. doi 10.1080/15459624.2014.933960.

22. Intergovernmental Panel on Climate Change. *Climate Change 2014: Impacts, Adaptation, and Vulnerability*. New York: Cambridge University Press, 2014. http://www.ipcc.ch/report/ar5/wg2 (accessed August 1, 2014).

23. Phillips NG, Ackley R, Crosson ER, et al. Mapping urban pipeline leaks: methane leaks across Boston. *Environmental Pollution* 173 (2013): 1–4. doi 10.1016/j.envpol.2012.11.003.

24. Jackson RB, Down A, Phillips NG, et al. Natural gas pipeline leaks across Washington, DC. *Environmental Science & Technology* 48, no. 3 (2014): 2051–2058. doi 10.1021/es404474x.

25. U.S. Environmental Protection Agency. *Oil and Natural Gas Air Pollution Standards*. http://www.epa.gov/airquality/oilandgas (accessed August 1, 2014).

26. International Energy Agency. *Golden Rules for a Golden Age of Gas*. 2012. http://www.worldenergyoutlook.org/goldenrules (accessed August 1, 2014).

27. U.S. Environmental Protection Agency. *Reduced Emissions Completions for Hydraulically Fractured Natural Gas Wells*. 2011. http://www.epa.gov/gasstar/documents/reduced_emissions_completions.pdf (accessed August 1, 2014).

28. *Oil and Natural Gas Sector: New Source Performance Standards and National Emission Standards for Hazardous Air Pollutants Reviews* (EPA-HQ-OAR-2010-0505; FRL-9665-1). U.S. Environmental Protection Agency, 2012. http://www.epa.gov/airquality/oilandgas/pdfs/20120417finalrule.pdf (accessed August 1, 2014).

29. Hendler A, Nunn J, Lundeen J. *VOC Emissions from Oil and Condensate Storage Tanks. Final Report*. Woodlands, TX: Texas Environmental Research Consortium, 2006. http://startelegram.typepad.com/barnett_shale/files/environmental_h051cfinalreport.pdf (accessed August 1, 2014).

30. Martin R, Moore K, Mansfield M, et al. *Final Report: Uinta Basin Winter Ozone and Air Quality Study December 2010–March 2011*. Vernal: Energy Dynamics Laboratory, Utah State University Research Foundation, 2011. http://rd.usu.edu/files/uploads/ubos_2010-11_final_report.pdf (accessed August 1, 2014).

31. Carter WPL, Seinfeld JH. Winter ozone formation and VOC incremental reactivities in the Upper Green River Basin of Wyoming. *Atmospheric Environment* 50 (2012): 255–266. doi doi.org/10.1016/j.atmosenv.2011.12.025.

32. Swarthout RF, Russo RS, Zhou Y, et al. Volatile organic compound distributions during the NACHTT campaign at the Boulder Atmospheric Observatory: influence of urban and natural gas sources. *Journal of Geophysical Research: Atmospheres* 118, no. 18 (2013): 10614–10,637. doi 10.1002/jgrd.50722.

33. Pétron G, Frost G, Miller BR, et al. Hydrocarbon emissions characterization in the Colorado Front Range: a pilot study. *Journal of Geophysical Research: Atmospheres* 117, no. D4 (2012), D04304. doi 10.1029/2011JD016360.

34. Helmig D, Thompson C, Evans J, et al. Highly elevated atmospheric levels of volatile organic compounds in the Uinta Basin, Utah. *Environmental Science and Technology* 48, no. 9 (2014): 4707–4715. doi 10.1021/es405046r.

35. Zielinska B, Fujita B, Campbell B. *Monitoring of Emissions from Barnett Shale Natural Gas Production Facilities for Population Exposure Assessment*. Houston,

TX: Desert Research Institute. https://sph.uth.edu/mleland/attachments/Barnett%20Shale%20Study%20Final%20Report.pdf (accessed August 1, 2014).

36. Frazier A. *Analysis of Data Obtained for the Garfield County Air Toxics Study, Summer 2008*. http://www.garfield-county.com/air-quality/documents/airquality/2008_Targeted_Oil_and_Gas_Monitoring_Report.pdf (accessed August 1, 2014).

37. Honeycutt M. *Air Quality Impacts of Natural Gas Operations in Texas*. Texas Commission on Environmental Quality, 2012. http://www.iom.edu/~/media/Files/Activity%20Files/Environment/EnvironmentalHealthRT/2012-04-30/Honeycutt.pdf (accessed August 1, 2014).

38. *Southwestern Pennsylvania Marcellus Shale Short-Term Ambient Air Sampling Report*. Bureau of Air Quality, Pennsylvania Department of Environmental Protection, 2010. http://www.dep.state.pa.us/dep/deputate/airwaste/aq/aqm/docs/Marcellus_SW_11-01-10.pdf (accessed August 1, 2014).

39. Field et al., Air quality concerns, op. cit.

40. Colborn T, Schultz K, Herrick L, et al. An exploratory study of air quality near natural gas operations. *Human and Ecological Risk Assessment: An International Journal* 20 (2013): 86–105. doi 10.1080/10807039.2012.749447.

41. Hendler et al., *VOC Emissions*, op. cit.

42. Esswein et al., Evaluation of some potential chemical exposure risks, op. cit.

43. Glass DC, Gray CN, Jolley DJ, et al. Leukemia risk associated with low-level benzene exposure. *Epidemiology* 14, no. 5 (2003): 569–577.

44. Kirkeleit J, Riise T, Bratveit M, et al. Increased risk of acute myelogenous leukemia and multiple myeloma in a historical cohort of upstream petroleum workers exposed to crude oil. *Cancer Causes and Control* 19, no. 1 (2008): 13–23. doi 10.1007/s10552-007-9065-x.

45. White N, teWaterNaude J, van der Walt A, et al. Meteorologically estimated exposure but not distance predicts asthma symptoms in schoolchildren in the environs of a petrochemical refinery: a cross-sectional study. *Environmental Health* 8, no. 1 (2009): 45. doi 10.1186/1476-069X-8-45.

46. Agency for Toxic Substances and Disease Registry (ATSDR). *Toxicological Profile for Benzene*. Atlanta, GA: U.S. Department of Health and Human Services, 2007. http://www.atsdr.cdc.gov/toxprofiles/tp.asp?id=40&tid=14 (accessed August 1, 2014).

47. Goldstein BD. Benzene as a cause of lymphoproliferative disorders. *Chemico-Biological Interactions* 184, no. 1–2 (2010): 147–150. doi.org/10.1016/j.cbi.2009.12.021.

48. Steinzor N, Subra W, Sumi L. Investigating links between shale gas development and health impacts through a community survey project in Pennsylvania. *New Solutions* 23, no. 1 (2013): 55-83. doi 10.2190/NS.23.1.e.

49. Guidotti TL. Hydrogen sulfide advances in understanding human toxicity. *International Journal of Toxicology* 29, no. 6 (2010): 569–581; doi 10.1177/1091581810384882.

50. Goldstein BD, Kriesky J, Pavliakova B. Missing from the table: role of the environmental public health community in governmental advisory commissions related to Marcellus Shale drilling. *Environmental Health Perspectives* 120, no. 4 (2012): 483–486; doi 10.1289/ehp.1104594.

51. Witter RZ, McKenzie LM, Towle M, et al. *Health impact assessment for Battlement Mesa, Garfield County Colorado*. Colorado School of Public Health, 2011. http://www.garfield-county.com/public-health/documents/1%20%20%20Complete%20HIA%20without%20Appendix%20D.pdf (accessed August 1, 2014).

52. Saberi P. Navigating medical issues in shale territory. *New Solutions* 23, no. 1 (2013): 209–221. doi.org/10.2190/NS.23.1.m

53. ATSDR. *Toxicological Profile for Xylenes*. Atlanta, GA: U.S Department of Health and Human Services, 2007. http://www.atsdr.cdc.gov/toxprofiles/tp.asp?id=296&tid=53 (accessed August 1, 2014).

54. Office of Pollution Prevention and Toxics. *Chemicals in the environment: 1,2,4-trimethylbenzene* (C.A.S. No. 95-63-6; EPA 749-F-94-022). Washington, DC: U.S. Environmental Protection Agency, 1994. http://www.epa.gov/chemfact/f_trimet.txt (accessed August 1, 2014).

55. ATSDR. *Toxicological Profile for Xylenes*, op. cit.

56. McKenzie LM, Guo R, Witter RZ, et al. Birth outcomes and maternal residential proximity to natural gas development in rural Colorado. *Environmental Health Perspectives* 122 (2014): 412–417. doi 10.1289/ehp.1306722.

57. Fryzek J, Pastula S, Jiang X, et al. Childhood cancer incidence in Pennsylvania counties in relation to living in counties with hydraulic fracturing sites. *Journal of Occupational and Environmental Medicine* 55, no. 7 (2013): 796–801. doi 10.1097/JOM.0b013e318289ee02.

58. Goldstein BD, Malone S. Obfuscation does not provide comfort. *Journal of Occupational and Environmental Medicine* 55, no. 11 (2013): 1376–1378. doi 10.1097/JOM.0000000000000014.

59. Esswein EJ, Breitenstein M, Snawder J, et al. Occupational exposures to respirable crystalline silica during hydraulic fracturing. *Journal of Occupational and Environmental Hygiene* 10, no. 7 (2013): 347-56. doi 10.1080/15459624.2013.788352.

60. Worker exposure to silica during hydraulic fracturing hazard alert. Washington, DC: National Institute for Occupational Safety and Health, 2012. http://www.osha.gov/dts/hazardalerts/hydraulic_frac_hazard_alert.html (accessed August 1, 2014).

61. Occupational Safety and Health Administration. *OSHA Fact Sheet: OSHA's Proposed Crystalline Silica Rule: Overview*. https://www.osha.gov/silica/factsheets/OSHA_FS-3683_Silica_Overview.html (accessed July 15, 2014).

62. Occupational Safety and Health Administration. *Crystalline Silica Rulemaking*. https://www.osha.gov/silica/index.html (accessed July 14, 2014).

63. Litovitz A, Curtright A, Abramzon S, et al. Estimation of regional air-quality damages from Marcellus Shale natural gas extraction in Pennsylvania. *Environmental Research Letters* 8, no. 1 (2013): 14–17. doi 10.1088/1748-9326/8/1/014017.

64. National Toxicology Program, *Diesel Exhaust Particulates*, op. cit.

65. Rich et al., An exploratory study of air emissions, op. cit.

66. U.S. Environmental Protection Agency. *Nonroad Diesel Engines*. http://www.epa.gov/otaq/nonroad-diesel.htm (accessed July 22, 2014).

67. Integrated Risk Information System, U.S. Environmental Protection Agency. *Diesel engine exhaust*. 2003. http://www.epa.gov/iris/subst/0642.htm (accessed July 22, 2014).

68. International Agency for Research on Cancer. *Monographs on the Evaluation of Carcinogenic Risks to Humans*. Geneva: World Health Organization, 2013. http://monographs.iarc.fr/ENG/Classification/ClassificationsAlphaOrder.pdf (accessed July 15, 2014).

69. Hart JE, Rimm EB, Rexrode KM, et al. Changes in traffic exposure and the risk of incident myocardial infarction and all-cause mortality. *Epidemiology* 24, no. 5 (2013): 734–742. doi 10.1097/EDE.0b013e31829d5dae.

70. Hart JE, Laden F, Schenker MB, et al. Chronic obstructive pulmonary disease mortality in diesel-exposed railroad workers. *Environmental Health Perspectives* 114, no. 7 (2006): 1013–1017. doi 10.2307/3651770.

71. Hesterberg TW, Long CM, Bunn WB, et al. Non-cancer health effects of diesel exhaust: a critical assessment of recent human and animal toxicological literature. *Critical Reviews in Toxicology* 39, no. 3 (2009): 195–227. doi 10.1080/10408440802220603.

72. U.S. Environmental Protection Agency. *Oil and Gas Production Wastes*. http://www.epa.gov/radiation/tenorm/oilandgas.html (accessed August 4, 2014).

73. Nicoll G. Radiation sources in natural gas well activities. *Occupational Health and Safety* 81, no. 10 (2012): 22, 24, 26.

74. U.S. Environmental Protection Agency, *Oil and Gas Production Wastes*, op. cit.

75. Occupational Safety and Health Administration. *OSHA Fact Sheet: Hydrogen Sulfide*. 2005. https://www.osha.gov/OshDoc/data_Hurricane_Facts/hydrogen_sulfide_fact.pdf (accessed August 4, 2014).

76. Witter RZ, Tenney L, Clark S, et al. Occupational exposures in the oil and gas extraction industry: state of the science and research recommendations. *American Journal of Industrial Medicine* 57, no. 7 (2014): 847–856. doi 10.1002/ajim.22316.

77. Occupational Safety and Health Administration, *OSHA Fact Sheet: Hydrogen Sulfide*, op. cit.

78. U.S. Environmental Protection Agency. *Ground-level Ozone: Basic Information*. http://www.epa.gov/air/ozonepollution/basic.html (accessed July 5, 2014).

79. Schnell RC, Oltmans SJ, Neely RR, et al. Rapid photochemical production of ozone at high concentrations in a rural site during winter. *Nature Geoscience* 2, no. 2 (2009): 120–122. doi.org/10.1038/ngeo415.

80. Katzenstein AS, Doezema LA, Simpson IJ, et al. Extensive regional atmospheric hydrocarbon pollution in the southwestern United States. *Proceedings of the National Academy of Sciences* 100, no. 21 (2003): 11975–11979. doi 10.1073/pnas.1635258100.

81. Gilman JB, Lerner BM, Kuster WC, et al. Source signature of volatile organic compounds from oil and natural gas operations in northeastern Colorado. *Environmental Science and Technology* 47, no. 3 (2013): 1297–1305. doi 10.1021/es304119a.

82. Kemball-Cook S, Bar-Ilan A, Grant J, et al. Ozone impacts of natural gas development in the Haynesville Shale. *Environmental Science and Technology* 44, no. 24 (2010): 9357–9363. doi 10.1021/es1021137.

83. Utah Department of Environmental Quality. *Final Report: 2012 Uintah Basin Winter Ozone & Air Study. Uintah Basin Winter Ozone Study*. 2013. http://rd.usu.edu/files/uploads/ubos_2011-12_final_report.pdf (accessed August 14, 2014).

84. Edwards PM, Young CJ, Aikin K, et al. Ozone photochemistry in an oil and natural gas extraction region during winter: simulations of a snow-free season in the Uintah Basin, Utah. *Atmospheric Chemistry and Physics Discussions* 13, no. 3 (2013): 7503–7552. doi 10.5194/acpd-13-7503-2013.

85. Air quality criteria for ozone and related photochemicaloxidants (EPA/600/R-05/004aF-cF). Research Triangle Park, NC: Office of Research and Development, National Center for Environmental Assessment, U.S. Environmental Protection Agency, 2006. http://cfpub.epa.gov/ncea/cfm/recordisplay.cfm?deid=149923 (accessed August 14, 2014).

86. Pride K, Peel J, Robinson B, et al. Associations of short-term exposure to ozone and respiratory outpatient clinic visits—Sublette County, Wyoming, 2008–2011. Cheyenne, WY: State of Wyoming Department of Health, 2013. https://cste.confex.com/cste/2013/webprogram/Paper1219.html (accessed August 4, 2014).

87. Allen et al., Measurements of methane emissions, op. cit.

Chapter 6

Implications of Unconventional Gas Extraction on Climate Change: A Global Perspective

Philip L. Staddon and Michael H. Depledge

CAUSES AND CONSEQUENCES OF CLIMATE CHANGE

Climate change is one of the greatest 21st-century challenges that societies must address in the coming years.[1] The evidence is overwhelming that climate change is being driven primarily by human activities, including the release into the atmosphere of greenhouse gases, carbon dioxide (CO_2), and methane (CH_4), as a result of burning fossil fuels and land use changes.[2] Evidence dating from the Industrial Revolution clearly shows a link between the build-up of CO_2 concentration in the atmosphere and the temperature of the planet.[3] The build-up of these gases has had a profound effect on the earth's temperature over this time period, yet industrialized and developing nations have not sufficiently moderated the quantity of greenhouse gases emitted into the atmosphere.

Even in the extremely unlikely event that nations could or would quickly curtail their use of fossil fuels and reduce discharges of CO_2 and other gases into the atmosphere, because of the time lags involved in the global carbon cycle, it is estimated that by 2100, there will be an increase in the global average temperature by at least 1°C, with greater warming at higher latitudes.[4] The latest projections for climate change under varying demographic and economic scenarios (ranging from the overly optimistic to the laissez-faire of business-as-usual) put the increase in global temperature by 2100 between 1 and 5°C, with 2 to 4°C being the most likely, depending on future energy choices. Most scientists believe that a 2°C increase will result in a dangerous change in the global climate.[5]

Climate change, often also referred to as "global warming," undoubtedly will contribute to changes in environmental living conditions,[6,7] and change will affect different regions more or less intensely depending in part on the resistance and resilience of local ecosystems and local weather patterns. As the climate warms, precipitation patterns change, and extreme weather events become more frequent, not only affecting all sectors of the economy and society but also leading to adverse effects on infrastructure, food production (agriculture and fisheries), business activity, ecosystem services, and human health.[8] For example, the world has already experienced unusually severe weather events including the 2003 European heat wave, Hurricane Katrina in the United States in 2005, Typhoon Haiyan in southeast Asia in 2013, flooding in central England in 2014, the polar vortex (extreme cold) in many parts of the United States during the winter of 2014, and the ongoing drought in California. It is not possible to say with absolute certainty that any one of these extreme weather events is directly attributable to climate change; however, the increasing frequency and intensity of such events, as well as the frequency with which long-standing weather records are being broken, attest to a change in climate.

Despite the evidence of climate change, there are some who continue to question its existence. For many, climate change and global warming are things that will have an impact far into the future. However, it is hard to deny that human-induced climate change is already evident all around us. Many plants have changed the timing of their phenology (e.g., bud burst, flowering). In England, for example, the timing of seasonal events for species has advanced by up to 3 weeks in little more than a few decades.[9] Some plant and animal species have already been noted to extend their range northward or to higher altitudes in mountainous areas. Range shift applies to all habitats and in many instances is more marked in the marine environment, possibly because of the free movement of organisms in a fluid environment. These changes will have extensive repercussions for many ecosystems, including agricultural settings.[10]

As different species acclimatize to new climatic and environmental conditions at different rates (e.g., human encroachment, land-use change, fertilizer and pollutant loads), there will be an emergence of new assemblages of species with associated alterations in ecosystem functioning and "services."[11] Problematic species (from a human point of view) including "invasives" and "aliens" (e.g., species new to an area that cause degradation and loss of diversity to the local ecosystems and/or nefarious impacts on agricultural, forestry, and fishery activities, leading to direct or indirect costs to society) take advantage of the changing environmental conditions and the pressures exerted on established ecosystems, some of which will become increasingly ecologically unstable, allowing easy entry to "nuisance" species.[12] For example, there is already evidence of the spread of plant and animal pests and diseases (e.g., bluetongue disease spread by midges in sheep) and changes in species that have an impact on human health (e.g., malaria spread by mosquitos expanding their ranges to higher latitudes and altitudes; the spread of the allergenic pollen producing plant *Ambrosia* across Europe).

The effect of climate change on human health and disease is of particular concern.[13] Direct effects on health undoubtedly will follow heat waves, cold spells, flooding, storms, and other extreme weather events.[14] Indirect effects attributable to climate change undoubtedly will have an impact on water quality, food quality and security, and socioeconomic conditions, among other factors.[15] Of note, in 2000, the World Health Organization (WHO) identified climate change as a key health risk.[16] In 2000, climate change was estimated to have caused the loss of 5.5 million disability adjusted life years (DALYs) globally, which included a significant proportion attributable to diarrhea, malaria, and malnutrition.[17] The situation has worsened over the ensuing 15 years. Perhaps not surprisingly, children are disproportionately affected.[18] Climate change is estimated to account for approximately 3% of diarrhea, 3% of malaria, and 4% of dengue cases globally. A rise in global temperature of only 1°C would increase deaths from diarrhea, malaria, and malnutrition by 300,000. A 2°C rise in global temperature could result in 40 to 60 million more people being exposed to malaria in Africa, with millions more at risk of malnutrition and starvation because of declining crop yields.[19] Other estimates predict that by 2080, climate change will result in between 260 and 320 million people being exposed to malaria and approximately 2.5 billion exposed to dengue fever.[20]

GLOBAL GREENHOUSE GAS EMISSIONS—PAST AND FUTURE

A greenhouse gas (GHG) is a gas that both absorbs and emits radiation in the infrared range, commonly called thermal radiation or heat. When present in the atmosphere, these gases trap radiation in the form of heat, causing a warming process called the greenhouse effect. Since the beginning of the Industrial Revolution (taken as the year 1750), the burning of fossil fuels and extensive clearing of native forests has contributed to a 40% increase in the atmospheric concentration of carbon dioxide, from 280 to 392.6 parts per million (ppm). Currently, global energy consumption is predominantly based on fossil fuels, which account for over 80% of the CO_2 emissions of which 46% is attributed to coal, 34% to oil and 20% to gas.[21] China accounts for 26.9% of global annual CO_2 emissions from fossil fuels, and the United States, until recently the world's biggest polluter, accounts for 14.5%. India (6.3%), Russia (5.5%), Japan (3.6%), and Germany (2.1%) are also significant emitters of fossil fuel CO_2. Africa and Latin America account for a mere 3.6% and 5.2%, respectively.[22]

Historically, cumulative emissions dating back to the Industrial Revolution also show the unequal responsibilities of regions in generating climate change: 72% are attributable to the developed world, of which the U.S. contribution is estimated to be 26% and China's is 10.7% despite a much larger share of the world's population.[23] It should be recognized when assigning the present share of responsibility for global GHG emissions that consumers in developed countries drive a large part of the emissions by producers in rapidly developing countries. In fact, the rate of increase in CO_2 emissions from fossil fuels has accelerated in the past decade, principally because of the emergence of China and other fast-developing nations (e.g., India, Indonesia, and Brazil) as new regional and global economic powers.

In other words, consumers in rich countries are responsible for emissions in poor countries producing the goods that consumers in rich countries buy. In effect, rich countries are currently exporting GHG emissions to lower- and middle-income countries from where they import manufactured goods. Or, put another way, rich countries maintain their standard of living with nominally lower GHG emissions by transferring these emissions to poor countries. This situation highlights a long-debated question of whether GHG emissions should be measured at the point of consumption rather than at the point of production. One could argue that GHG should be measured at the level of the consumer, in which case the United States, for example, would see its share of responsibility for GHG emissions vastly increase.

Despite the myriad international agreements and promises by governments around the world, globally CO_2 emissions from fossil fuel combustion have increased from 19 gigatons (Gt) CO_2 yr^{-1} in 1980 to 30 Gt CO_2 yr^{-1} in 2010.[24] The growth rate of global emissions of fossil fuels has continued to accelerate from 1.5% per year between 1980 and 2000 to 3% per year between 2000 and 2012 primarily due to a major increase in coal use.[25] Over the same time period, methane (CH_4) and nitrous oxide (N_2O) emissions associated with human activity, especially in the agricultural and the energy sectors, have remained stable at ca. 9 Gt and 3 Gt CO_2-eq yr^{-1}, respectively. Despite this gloomy situation, there is some positive action happening. There is evidence that ozone-depleting substances such as chlorofluorocarbon (CFCs) and hydro-chlorofluorocarbons (HCFCs) have nearly been eliminated, which is good news for mitigating the ozone hole.[26]

Given their substantial role in climate change, it is important to include fossil fuel reserves in the climate change debate. Globally, total CO_2 emissions are relatively small compared with the total CO_2 that would be emitted should the estimated reserves of all fossil fuels be burned. However, the picture varies with different fossil fuel sources. For conventional oil and gas, it is estimated that 60% to 80% of total estimated reserves have yet to be extracted. Vast global reserves of unconventional oil and gas will double the reserves for oil and multiply gas reserves tenfold. Readily recoverable coal reserves are approximately threefold higher than resources already extracted and total estimated reserves are more than 10,000 Gt C, which is more than 50 times higher than the volume of all coal already burned.[27]

At what point recoverable fossil fuel reserves will be depleted is open for debate. Optimistic time scales (from a climate change view point) set 35 and 37 years for oil and gas, respectively, and 107 years for coal.[28,29] Under this scenario, CO_2 emissions would lead to a CO_2 concentration in the atmosphere that would be approximately 480 ppm, which would keep global warming below 2°C.[30] However, much of the so called "unrecoverable" fossil fuel reserve will in fact become available as technology develops, which helps explain why the ratio of reserve-to-production for oil has fluctuated between 20 to 40 years for most of the 20th century.[31]

Within the past decade, there has been increasing effort focused on exploiting previously untouched sources of fossil fuel. New technologies and improvements to existing ones have vastly increased the amount of fossil fuel that is economically recoverable, especially the extraction of oil from tar sands, natural gas trapped in shale, and coal by surface mining. On the basis of current estimates (and current technologies), unconventional gas could represent 40% of recoverable gas resources globally, with shale gas accounting for 27%.[32] In addition, increasing effort is being focused on exploiting previously untouched sources of fossil fuel by developing technologies for deep ocean oil and gas drilling and for exploitation of sea bed methane hydrates.

Hansen et al.[33] assessed climate impacts of global warming using ongoing observations and paleoclimate data. Despite evidence of the effects of high fossil fuel emissions in climate change, governments continue to allow and even encourage pursuit of ever more fossil fuels. For example, future cumulative CO_2 emissions from global existing infrastructure over the next 50 years range between 282 and 701 Gt of CO_2 resulting in stabilization of atmospheric CO_2 concentration below 430 ppm and a warming of 1.1°C to 1.4°C above the preindustrial level.[34] This estimate assumes that no future fossil fuel infrastructure is produced, thus highlighting the unlikely possibility of remaining below a 2°C rise in global temperature. Furthermore, it has been estimated that by 2100, China's future CO_2 emissions will substantially increase from the current ca. 4 Gt CO_2 yr^{-1} to ca. 10

Gt CO_2 yr^{-1}.[35] This would be equivalent to cumulative emissions of 353 Gt CO_2 between 2010 and 2060, and 707 Gt CO_2 by 2100. Thus, keeping below the 2°C threshold rise is unlikely under these circumstances unless carbon capture is implemented on an industrial scale.

Given all this, it appears extremely unlikely that any abatement in climate change can be avoided. As Rogner[36] noted, neither hydrocarbon resource availability nor costs are likely to become forces that would help encourage the global energy system to reduce the use of fossil fuel during the next century.

HOW WILL HYDRAULIC FRACTURING OF SHALE GAS (FRACKING) AFFECT FUTURE GHG EMISSIONS?

Several countries around the world have vast reserves of recoverable shale gas, many times greater than their conventional gas reserves; examples of these countries are China, Argentina, France, Canada, and the United States.[37] The United States is the acknowledged global leader in shale gas extraction having many huge shale formations including the Bakken Formation and the Marcellus Shale, each encompassing hundreds of thousands of square miles. In 2013, approximately 30% of U.S. gas production came from shale gas; it is estimated that the figure will rise to 45% by 2035.[38] If all these unconventional gas resources were to be mobilized, the resulting CO_2 and CH_4 emissions would add to climatic forcing and further exacerbate climate change.

Shale gas is portrayed as being "environmentally friendly" by the oil and gas industry as well as by governments that continue to support fossil fuel use. Proponents argue that CO_2 emissions from shale gas are lower than for coal and oil and that shale gas is a logical transition fuel to CO_2 emission-free alternative energy sources.[39] However, there is a fallacy in this position.[40] Shale gas actually represents an *additional* source of fossil fuel GHG emissions. Shale gas may actually delay rather than speed up the transition to renewable and low-carbon energy sources.

Furthermore, there is concern that the shale gas GHG footprint is much larger than is widely recognized. Unconventional extraction of shale gas releases large quantities of methane into the atmosphere. It is estimated that between 3.6% and 7.9% of the methane produced at shale-gas wells escapes into the atmosphere.[41,42] This is at least 30% more than the methane losses from conventional gas production. Others estimate that methane emissions have been greatly underestimated in local and regional inventories of GHG emissions.[43] The methane escaping during shale gas extraction and the release of CO_2 during gas end use mean that the carbon footprint (CO_2 + CH_4) of shale gas is significantly greater than that of conventional gas and is actually very similar to coal when compared over the long term (e.g., 100 years). It should be noted that there is some debate regarding the veracity of these calculations.[44,45] However, other independent studies[46] have reached similar conclusions to Howarth et al., and the U.S. Environmental Protection Agency's latest estimates of fugitive methane emissions from shale gas wells also is supportive.[47]

Liquefied natural gas (LNG) is another source of GHG emissions. LNG is used as a means of facilitating transport and export of gas in the absence of gas pipelines, especially for transocean shipments. Both the liquefaction and regasification processes use large amounts of energy, thereby releasing further GHG into the atmosphere. The life cycle GHG emissions of LNG are 18% to 20% higher than for domestically consumed gas.[48] This point is particularly relevant as increasing amounts of shale gas from the United States and Canada are being exported as LNG. Many LNG import terminals (regasification) have been refurbished to serve as export terminals (liquefaction). As a result of increased extraction, the United States has become a net gas exporter, much of it in the form of LNG exports.[49]

ENVIRONMENTAL AND HEALTH IMPACTS OF ACCELERATED CLIMATE CHANGE

The speed and magnitude of climate change will determine the strength and intensity of the various damaging effects on the environment and on health and well-being. Accelerated climate change is expected to (1) bring forward the point at which adverse impacts on health become more serious, (2) shorten the time available to implement adaptation strategies to counter these impacts, (3) limit the chance of successful acclimatization to a changing climate by natural ecosystems, (4) lead to greater weather volatility (as the disruption of the weather systems occurs more rapidly), and (5) lead to the climate stabilizing at a warmer global temperature than would have been reached by slower climate change (as more time would have been available to limit GHG emissions). Indeed, accelerated climate change makes the task of guaranteeing societal resilience to extreme weather ever more challenging.

Adverse impacts on agricultural food production include those caused by insufficient or unpredictable water availability, soil erosion and decreasing soil quality (many soils are now of low quality because of poor management), crop diseases and pests (favored in the new climatic conditions and/or attacking stressed crops), and animal diseases (especially if animals are already otherwise stressed). There are numerous examples of recent emerging/spreading diseases in both plants and animals: Schmallenberg virus disease in sheep, bluetongue virus disease in ruminants (especially sheep, but also cattle, goats, and others), ash dieback fungal disease caused by an ascomycete, and many crop fungal diseases.[50]

There is a strong likelihood that accelerated climate change will precipitate increased weather variability and volatility leading to an increase in major public health challenges that could adversely affect individuals' physical, physiological, and psychological status.[51] This is particularly so given changing demographics (notably, aging of the population) and migration to urban areas where, for example, the effects of heat waves are exacerbated by the urban heat island effect.[52] Rising water levels due to global warming also pose a challenge. Twenty-three of the 30 largest cities in the world are at sea level, including Guangzhou, Shanghai, Mumbai, Chittagong, Alexandria, Lagos, London, and New York.

Droughts, which are likely to become much more frequent and intense, will be felt hardest in the poorest areas of Africa and will severely compromise water and food supplies around the world. Niger and Sudan in the Sahel, Burundi, and Tanzania in East Africa and Malawi and Zimbabwe in southern Africa are likely to be the most affected by drought. Prolonged periods of drought have the potential to lead to water wars.[53] Water wars have already been noted in the Sudan and in Mali, leading to mass migration.[54]

Climate change also contributes to water quality decline and waterborne diseases, marine risks (e.g., jellyfish swarms), vectorborne diseases (via mosquitoes, tse tse flies, ticks), and hayfever and asthma (via pollen).[55,56] Of particular worry is the spread of many warm-climate diseases transmitted by mosquitoes and tse tse flies to higher latitudes and altitudes.[57] These diseases, which include malaria, dengue, chikungunya, and sleeping sickness, affect hundreds of millions of people each year.

CLIMATE CHANGE MITIGATION AND ADAPTATION

There are two main policy responses to climate change: mitigation and adaptation. Mitigation addresses the root causes by reducing GHG emissions. Climate change mitigation refers to efforts to reduce or prevent emission of GHG. Examples of mitigation include switching to low-carbon energy sources, such as renewable and nuclear energy. An important point to underline is that mitigation aims to control a global phenomenon and therefore requires international cooperation for a successful outcome. Adaptation seeks to lower the risks posed by the consequences of climatic changes. Both approaches will be necessary because even if emissions are dramatically decreased in the next decade (an unlikely scenario), adaptation will still be needed to deal with the global changes that have already been set in motion.

If one accepts the thesis presented here, one would conclude that shale gas expansion is inconsistent with climate change mitigation primarily because the large GHG footprint of shale gas undercuts the logic of its use as a bridging fuel over coming decades. If the goal is to reduce global warming, lowering GHG emissions will require the use of less, not more, fossil fuels, making the search for new sources of fossil fuels seem particularly perverse. One caveat here is that if it can be properly regulated, a switch from coal to shale gas could be beneficial in the short term to limit local air pollution, a major cause of early mortality worldwide, especially in fast-growing cities in rapidly industrializing nations such as China and India. Over the long term, reliance on shale gas would be of limited use in mitigating climate change. Nonetheless, it is worth highlighting that climate change mitigation efforts have many health and environmental co-benefits that can be achieved now by decreasing reliance on fossil fuels, such as decreasing respiratory illnesses by enforcing clean air protocols.[58]

In summary, contrary to mitigation, adaptation must be delineated and implemented at the local level. Shale gas extraction makes the need for climate change adaptation ever more urgent. The evidence to date is that mitigation targets will (1) not be met by the international community and (2) not be sufficient to avert significant climate change. The key to counter the arguments that we must wait for "certainty" is to (1) focus on building adaptive capacity and improving resilience to climatic events; (2) ensure a flexible approach to adaptation, which can be fully responsive to ongoing changes; and (3) improve current preparedness for extreme weather events (e.g., flooding, storms, heat waves, cold spells), which will be beneficial immediately as well as in the future.

CONCLUSION

Beyond any reasonable doubt, current climate change is caused by human activities and, in particular, the burning of fossil fuels. The adverse effects of climate change are already clearly evident on natural ecosystems, in agricultural production, and also on human societies. Yet governments around the world have either reluctantly agreed to limit greenhouse emissions, many at the same time as they actively engage in identifying and exploiting new reserves of fossil fuels including shale gas or have ignored all of the warnings of the potential harm of climate change to the earth.

Expanding fossil fuel extraction capabilities sends the message that climate change is not a major issue and that mitigation is of limited urgency; it starkly highlights the very low priority given to climate change mitigation by most governments and a lack of understanding by the wider public of climate change threats. The public perhaps perceives governments and large corporations downplaying the risks of climate change and concludes that the urgency is overstated. There is no reason or need to change our behavior. There is no urgency to tackle climate change or to invest in renewable energy and/or low- or no-carbon alternative energy sources. Moving to a sustainable future in such a context will be difficult.

If fossil fuels continue to be an acceptable source of energy over the next decades, then investment will continue to be pumped into fossil fuel technologies that provide a high return with relatively low economic risk. The major energy players all invest in alternative energies, but these investments are extremely small compared with the vast sums spent on fossil fuel exploration, research, and development. All of this significantly delays the switch away from fossil fuels, including shale gas, with the potential for major adverse consequences for the earth's climate and inhabitants. The energy industry promotes the idea of carbon capture and storage (CCS) as an answer to lower CO_2 emissions while permitting the continuing use of fossil fuels. However, to date there has been no large-scale demonstration that CCS would work and be safe [59,60]

There are claims that shale gas is "cleaner" than coal and could be used as a bridging fuel to a sustainable low-carbon future. However, unconventional shale gas development will significantly increase the cost of mitigation actions required to maintain a livable climate and will significantly increase the cost of adaptation actions required to support human societies in a less hospitable climate. The financial impacts in terms of damage to economies and to human health will be vast but will sadly be externalized and therefore borne by those societies affected by climate change, not by the business sector responsible for increased fossil fuel exploitation.

Carbon-neutral and low-carbon energy alternatives are available and should be further developed to limit the "need" for fossil fuels and to move to a sustainable future. What is at stake is the health and well-being of millions of people worldwide, and the sustainability of the earth's ecosystems.

NOTES

1. Intergovernmental Panel on Climate Change (IPCC). *Climate Change 2014: Impacts, Adaptation, and Vulnerability. Working Group II Contribution to the Fifth Assessment Report of the Intergovernmental Panel on Climate Change*. Cambridge, UK: Cambridge University Press, 2014.

2. IPCC. *Climate Change 2013: The Physical Science Basis. Working Group I Contribution to the Fifth Assessment Report of the Intergovernmental Panel on Climate Change*. Cambridge, UK: Cambridge University Press. 2013.

3. Arrhenius S. On the influence of carbonic acid in the air upon the temperature of the ground. *Philosophical Magazine and Journal of Science* 41 (1896): 237–275.

4. Matthews HD, Weaver AJ. Committed climate warming. *Nature Geoscience* 3 (2010): 142–143.

5. IPCC, *Climate Change 2013*, op. cit.

6. IPCC. *Managing the Risks of Extreme Events and Disasters to Advance Climate Change Adaptation. A Special Report of Working Groups I and II of the Intergovernmental Panel on Climate Change*. Cambridge, UK: Cambridge University Press, 2012.

7. Stern N. *The Economics of Climate Change. The Stern Review*. Cambridge UK: Cambridge University Press, 2007.

8. World Health Organization (WHO). *Protecting Health from Climate Change: Connecting Science, Policy and People*. Geneva: WHO Press, 2009.

9. Fitter AH, Fitter RSR. Rapid changes in flowering time in British plants. *Science* 296 (2002): 1689–1691.

10. IPCC, *Climate Change 2014*, op. cit.

11. Rohr JR, Dobson AP, Johnson PTJ, et al. Frontiers in climate change-disease research. *Trends in Ecology and Evolution* 26 (2011): 270–277.

12. Millennium Ecosystem Assessment (MEA). *Ecosystems and Human Well-being: Current State and Trends, Volume 1*. Washington, DC: Island Press, 2005.

13. Bell E. Readying health services for climate change: a policy framework for regional development. *American Journal of Public Health* 101 (2011): 804–813.

14. Costello A, Abbas M, Allen A, et al. Managing the health effects of climate change: Lancet and University College London Institute for Global Health Commission. *Lancet* 373 (2009): 1693–1733.

15. Godfray HCJ, Beddington JR, Crute IR, et al. Food security: the challenge of feeding 9 billion people. *Science* 327 (2010): 812–818.

16. WHO. *Climate Change and Human Health—Risks and Responses. Summary.* http://who.int.globalchange/summary/en/index12.html (accessed June 11, 2014).

17. Costello A, Maslin M, Montgomery H, et al. Global health and climate change: moving from denial and catastrophic fatalism to positive action. *Phil Trans R Soc A* 369 (2011): 1866–1882.

18. Sheffield PE, Landrigan PJ. Global climate change and children's health: threats and strategies for prevention. *Environmental Health Perspectives* 119 (2011): 291–298.

19. Stern, *The Economics of Climate Change*, op. cit.

20. Costello et al., Managing the health effects of climate change, op. cit.

21. Hansen J, Kharecha P, Sato M, et al. Assessing "dangerous climate change": required reduction of carbon emissions to protect young people, future generations and nature. *PLOS One* 8 (2013): e81648.

22. Ibid.

23. Ibid.

24. Montzka SA, Dlugokencky EJ, Butler JH. Non-CO_2 greenhouse gases and climate change. *Nature* 476 (2011): 43–50.

25. Hansen et al., Assessing "dangerous climate change," op. cit.

26. Montzka et al. Non-CO_2 greenhouse gases and climate change, op. cit.

27. Hansen et al., Assessing "dangerous climate change," op. cit.

28. Shafiee S, Topal E. When will fossil fuel reserves be diminished? *Energy Policy* 37 (2009): 181–189.

29. Höök M, Tang X. Depletion of fossil fuels and anthropogenic climate change—a review. *Energy Policy* 52 (2013): 797–809.

30. Chiari L, Zecca A. Constraints of fossil fuels depletion on global warming projections. *Energy Policy* 39 (2011): 5026–5034.

31. Rogner HH. An assessment of world hydrocarbon resources. *Annual Review of Energy and the Environment* 22 (1997): 217–262.

32. McGlade C, Speirs J, Sorrell S. Unconventional gas—a review of regional and global resource estimates. *Energy* 55 (2013): 571–584.

33. Hansen et al., Assessing "dangerous climate change," op. cit.

34. Davis SJ, Caldeira K, Matthews HD. Future CO_2 emissions and climate change from existing energy infrastructure. *Science* 329 (2010): 1330–1333.

35. Rout UK, Voβ A, Singh A, et al. Energy and emissions forecast of China over a long-time horizon. *Energy* 36 (2011): 1–11.

36. Rogner, An assessment of world hydrocarbon resources, op. cit.

37. Engelder T. Should fracking stop? No, it's too valuable. *Nature* 477 (2011): 271–275.

38. U.S. Energy Information Administration. *Outlook for Shale Gas and Tight Oil Development in the US*. http://www.eia.gov/pressroom/presentations/sieminski_04042013.pdf (accessed May 3, 2014).

39. Jenner S, Lamadrid AJ. Shale gas vs. coal: policy implications from environmental impact comparisons of shale gas, conventional gas, and coal on air, water, and land in the United States. *Energy Policy* 53 (2013): 442–453.

40. Stephenson E, Doukas A, Shaw K. Greenwashing gas: might a "transition fuel" label legitimize carbon-intensive natural gas development. *Energy Policy* 46 (2012): 452–459.

41. Howarth RW, Ingraffea A. Should fracking stop? Yes, it's too high risk. *Nature* 477 (2011): 271–273.

42. Howarth RW, Santoro R, Ingraffea A. Methane and the greenhouse-gas footprint of natural gas from shale formations. *Climatic Change* 106 (2011): 679–690. doi:10.1007/s10584-011-0061-5.

43. Pétron G, Frost G, Miller BR, et al. Hydrocarbon emissions characterisation in the Colorado Front Range: a pilot study. *Journal of Geophysical Research* 117 (2012): D04304. doi:10.1029/ 2011JD016360.

44. O'Sullivan F, Paltsev S. Shale gas production: potential versus actual greenhouse gas emissions. *Environmental Research Letters* 7 (2012): 044030.

45. Weber, CL, Clavin, C. Life cycle carbon footprint of shale gas: review of evidence and implications. *Environmental Science Technology* 46 (2012): 5688–5695.

46. Hultman N, Rebois D, Scholten M, Ramig C. The greenhouse impact of unconventional gas for electricity generation. *Environmental Research Letters* 6 (2011): 044008.

47. U.S. Environmental Protection Agency. *Inventory of US Greenhouse Gas Emissions and Sinks: 1990–2011*. Washington, DC: Author, 2013.

48. Stephenson E, Shaw K. A dilemma of abundance: governance challenges of reconciling shale gas development and climate change mitigation. *Sustainability* 5 (2013): 2210–2232.

49. Cueto-Felgueroso L, Juanes R. Forecasting long-term gas production from shale. *PNAS* 110 (2013): 19660–19661.

50. Fisher MC, Henk DA, Briggs CJ, et al. Emerging fungal threats to animal, plant and ecosystem health. *Nature* 484 (2012): 186–194.

51. Mason V, Andrews H, Upton D. The psychological impact of exposure to floods. *Psychology, Health, and Medicine* 15 (2010): 61–73.

52. Kim, HH. Urban heat island. *International Journal of Remote Sensing* 13 (1992): 2319–2336.

53. Godfray HCJ, Beddington JR, Crute IR et al. Food security: the challenge of feeding 9 billion people. *Science* 327 (2010): 812–818.

54. Gleick P. Water and conflict. *International Security* 18 (1993): 79–112.

55. McMichael C, Barnett J, McMichael AJ. An ill wind? Climate change, migration, and health. *Environmental Health Perspectives* 120 (2012): 646–654.

56. Smetacek V, Zingone A. Green and golden seaweed tides on the rise. *Nature* 504 (2013): 84–88.

57. Daszak P, Cunningham AA, Hyatt AD. Emerging infectious diseases of wildlife—threats to biodiversity and human health. *Science* 287 (2000): 443–449.

58. Costello et al. Global health and climate change, op. cit.

59. Tackling Climate Change-The Carbon Capture and Storage Association (CCSA). www.ccsassociation.org/why-ccs/tackling-climate change (accessed November 27, 2014).

60. Special Report on Carbon Dioxide Capture and Storage. IPCC. www.ipcc.ch/pdf/special-reports/SRCCS_wholerport.pdf (accessed November 27, 2014).

Chapter 7

Community Impacts of Shale-Based Energy Development: A Summary and Research Agenda

Kathryn J. Brasier and Matthew Filteau

INTRODUCTION

Over the past decade, the United States has experienced a rapid increase in the extraction of oil and natural gas from "unconventional" sources. These sources are termed unconventional because the natural gas is not concentrated within a reservoir but rather contained within tiny pockets and fissures within the geological formation.[1] The feasibility of extracting natural gas from these formations increased with the refinement of hydraulic fracturing and horizontal drilling techniques. Although both technologies had been used in the oil and gas industry for decades, the technologies were adapted and combined successfully to extract natural gas in the early 2000s from the Barnett Shale in the Dallas-Fort Worth region.[2,3]

Extraction of natural gas from "unconventional" geological formations, such as shale and tight sands, has been termed a "game-changer" in the energy system because of the ability to extract significant quantities of fossil fuels in the United States and many other parts of the world. The use of shale-based natural gas, because of the large domestic supply and low prices, is expected to grow significantly as a resource. The U.S. Energy Information Administration noted that U.S. shale gas production in 2005 was 0.75 trillion cubic feet (tfc) per year (4.1% of all gas produced in the United States); in 2012 that figure had grown to 9.7 tcf/year (40% of all gas produced), and is projected to increase to 19.8 tcf/year (53% of all gas produced) by 2040.[4]

As of 2013, natural gas provided just over one-quarter (26.6%) of the United States' energy needs, and continues to increase.[5] Natural gas is increasingly being used for electricity generation, particularly as coal-fired plants are retired. Projections suggest the continued importance of natural gas, and fossil fuels in general, to U.S. energy systems. However, public concerns about the social, environmental, and health concerns associated with the extraction, production, and consumption of energy suggest a need to understand the life-cycle impacts of energy sources on human and environmental systems. A few studies have begun to examine the life-cycle impacts (during extraction, transportation, consumption, and management of wastes) of, for example, the emissions of methane during natural gas extraction and transportation[6] and water use.[7] Within the social sciences, a growing body of literature has begun to examine the impacts on communities where natural gas extraction takes place. Much of the recent research chronicles community impacts in one region: the Marcellus Shale.

The Marcellus Shale is a natural-gas-bearing geological formation that lies beneath Pennsylvania, New York, Ohio, Maryland, and West Virginia; it has been described as the largest unconventional gas reserve in the United States and one of the largest worldwide.[8,9,10] The vast potential—and rising natural gas prices in the mid-2000s—spurred rapid development of unconventional natural gas in Pennsylvania beginning in 2005; by the end of 2013, approximately 7,400 unconventional gas wells had been drilled across the Commonwealth.[11] Pennsylvania moved from the seventh largest producer of natural gas in 2011 to the third largest producer in 2012, behind only Texas and Louisiana.[12] The Energy Information Administration projects that production from the Marcellus Shale will peak between 2022 and 2025, providing up to 39% of the natural gas needed to meet demand in markets east of the Mississippi River during that period—up from 16% in 2012.[13]

The rapid development of natural gas resources from the Marcellus Shale, and the prospects for continued growth in the region, has led to an emerging body of research about the social and community impacts of unconventional natural gas extraction. In this chapter, we summarize this social science literature, with a particular focus on recent research in the Marcellus Shale region. This formation has attracted a substantial amount of scholarly attention because of its location near major metropolitan centers and rapid, extensive drilling activity and development. We discuss the "boomtown model," which is the main theoretical approach used to understand social impacts on extractive communities. Throughout the chapter, we highlight how the Marcellus context is similar to, but differs from, the energy booms that scholars documented in earlier boomtown research in the 1970s and 1980s. We address how contemporary unconventional plays differ in development processes and in the contextual dissimilarities regarding spatial and temporal distribution of activity. We address how these conditions require scholars to forge new theoretical and empirical ground and provide a framework that encourages new questions.

ENERGY DEVELOPMENT AND "BOOMTOWNS"

Energy exploration and development begins when production companies contract specialty companies to complete specific tasks such as seismic testing, surveying, and leasing in preparation for drilling[14] in "hot spots" of natural resource extraction.[15] Each company brings its own equipment and workers because residents in these locations often do not have the technical training to perform these jobs. The influx of extra-local workers, and other migrants attracted to the potential economic opportunities, leads to a rapid population increase. This increase may provide a boost to the local economy, but it also spurs local inflation and labor shortages in some fields and stresses services, infrastructure, and local governments. This process has been dubbed the "boomtown phenomenon."

Social scientists used the term "boomtown" to refer to localities that experienced rapid natural resource development in isolated, rural communities in the intermountain West of the United States, primarily during the 1970s and 1980s. The effects were studied using a social disruption hypothesis, which specified that energy boomtowns undergo a period of generalized crisis and loss of traditional routines and attitudes during energy development.[16,17] This hypothesis is consistent with sociological theories about societies undergoing modernization (e.g., Durkheim, Tonnies), particularly rural communities experiencing processes of urbanization, and the loss of traditional culture, atomization, and disruption of social relationships that result.

Several early studies supported the social disruption hypothesis.[18,19,20] Rapid energy development in energy boomtowns was found to strain municipal services[21] and create social problems such as crime,[22] mental health issues,[23] and substance abuse.[24] Lack of adequate housing for the influx of workers is a critical problem, particularly in the early stages; new housing takes time to fund and build, forcing workers to live in temporary housing (e.g., hotels, RVs, mobile home parks). Existing residents, who often earn less than those in the energy industry, are priced out of the housing market and are pushed toward inadequate housing or out of the community altogether.[25]

Although some communities do experience economic growth, the economic benefits may not be captured by local economic systems, and the economic growth tends to be smaller than anticipated or projected.[26,27,28]

The validity of these findings, as well as the social disruption hypothesis, have been challenged and debated.[29] Wilkinson et al. reinforced the need for methodological rigor and research that accounted for the differential effects at the community and individual levels of analysis.[30] The question, Freudenburg argued, was no longer whether residents in boomtowns experienced negative social disruptions, but *which* communities and *which* residents experience change and to what degree.[31] In addition, these challenges pushed scholars to use longitudinal analyses to account for the differential effects from rapid development over time and across communities and individuals.[32,33]

Subsequent research shows how differential experiences result from (1) gender, (2) length of residence in the community, (3) age, (4) ability to receive direct economic benefits (such as a landowner or business owner), (5) stage of development, and (6) location of development (within or outside of the community). For example, women exhibited higher levels of community satisfaction than men during periods of energy booms.[34] Even though longtime boomtown residents reported the most severe decline in well-being,[35] they also maintained the highest levels of community satisfaction over the entire 25-year life course of energy development, whereas newcomers reported the lowest.[36] Newcomers who migrated to boomtowns during the energy boom reported fearing crime the most, and these fears apparently did not lessen with time, as they did with longtime residents.[37] These studies not only need to account for the study of individual characteristics within energy boomtowns but also how individuals experience change with time. One study found that boomtown residents' attitudes *recovered* 25 years after the boom and bust.[38,39,40]

Smith et al. contend that no two communities experience energy development equally.[41] In fact, population density, history with extractive industries, and type of extraction may affect how residents perceive energy development and social change in their communities. For example, Forsyth et al. found that longtime residents in Louisiana's coastal region perceived the effects from offshore oil development as either benign or positive.[42] This speaks to the importance of the location of development, type of extraction, and the history of energy extraction within a particular community. In Louisiana's oil-involved parishes, change in industrial activity is not associated with higher crime rates; in many locations where oil development persists, crime rates are lower.[43] These studies call for more attention on individual and community characteristics within energy boomtowns such as educational attainment, residential stability, and age as well as to the unique characteristics of different energy extraction processes.

Several authors point out that "booms" are usually followed by "busts." When the extracted resource becomes depleted or less profitable, a "cost-price squeeze" may occur.[44] It becomes more expensive to extract a less profitable resource, so production companies move their resources to other, more profitable locations. The employment opportunities offered by energy development begin to seep out of the community. Outmigration leads to decreasing revenue for local businesses and excess community infrastructure.[45] These boom–bust cycles are driven by a number of factors external to the local community, including demand, prices, changes in technology, organization of the extraction process, and global political forces.[46] Many rural communities that depend on natural resource extraction fall into continual boom and bust cycles, or what Freudenburg calls "addictive economies."[47] Longitudinal research by Brown et al. suggests that there is a stage that follows the boom and the bust, a "recovery stage," a period when community members adapt to and create new interpretations of their community.[48]

The boomtown model has provided a critical frame of reference for researchers studying the social and economic impacts of contemporary shale-based oil and gas development. In particular, studies of Marcellus Shale development have focused on the same set of critical impacts, including economic change, population change, housing impacts, institutional change (e.g., education, local government, human services, and health care services), crime, community relationships and conflict, and landscape change. In the next section, we review research on several topics that pertain to energy extraction in the Marcellus Shale region. We then discuss how, in the application of this model, researchers are beginning to question some of the fundamental assumptions behind the boomtown model and seek new theoretical lines of inquiry for describing and understanding the impacts of development within the region. We conclude with a research agenda based on our reading and understanding of the issue.

COMMUNITY IMPACTS IN THE MARCELLUS SHALE REGION: AN EMERGING LITERATURE

Research on the impacts of Marcellus Shale development has followed previous boomtown literature—that is, it describes changes across a number of community sectors. This work largely asks which community characteristics are changing and to what extent. Some of this research attempts to describe how these changes differ across space (by comparing communities within the region), over time (by comparing changes at multiple stages of development), and over individuals (by comparing experiences based on individual characteristics). However, as is noted subsequently, there is a need for additional work that describes how these differences are manifested, and what leads to those differences, within each topical area.

Economic Impacts

Economic impact studies in the Marcellus Shale region have largely focused on job growth and have resulted in widely varying estimates of jobs associated with Marcellus development. The highest estimates come from Considine, Watson, and Blumsack, who estimated that 44,098 direct, indirect, and induced jobs were created in Pennsylvania in 2009 and 139,889 in 2010.[49] However, these studies have been criticized by others.[50,51] In particular, Kelsey et al. estimate much lower job effects across the Commonwealth finding that only 23,884 jobs were generated in 2009.[52]

Jobs are not the only indicator of economic change. Kelsey et al. found that counties with high levels of Marcellus Shale activity also experience increased wages and income, substantially increased nonwage income (leasing and royalties), and greater levels of business activity.[53] The most significant effect is leasing and royalty income. Because only those who own their gas rights can benefit in this way, concerns are raised about the distribution of the benefits across residents in the region. Furthermore, to the extent that a significant proportion of gas rights are owned by nonresidents, leasing and royalty income may not be captured locally. Studies suggest that a substantial proportion of the increased wages are going to nonlocal (in- and out-of-state) workers, also raising questions about the extent to which communities within the Marcellus Shale are benefitting from the development.[54] Evidence of negative economic impacts is limited; Jacobson and Kelsey 2010 do not find higher municipal costs, and there is no documentation of negative effects on other sectors (such as tourism).[55] However, as Kelsey et al. note, the data that exist on benefits, such as job and income growth, are easier to track.[56]

Population

Using data from the decennial Census and the American Community Survey, McLaughlin et al. found no consistent association between population change and the level of Marcellus Shale development in Pennsylvania.[57] However, for a few counties in the northern tier of Pennsylvania, population trends changed from a loss in the early part of the decade, before Marcellus Shale development, to a gain in population in the latter part of the decade, during the period of high levels of Marcellus Shale development. These findings should be interpreted with some caution because itinerant workers are unlikely to be counted in the Census, which is likely to result in an underestimate of population change in this region.

Housing

A lack of affordable, quality housing has been identified as one of the most critical early impacts of Marcellus Shale development, especially in the most rural communities.[58,59] Many rural communities had relatively few housing options before the onset of Marcellus Shale development, and waves of short-term itinerant workers have filled temporary housing units (hotels, company-sponsored residential facilities, and campgrounds), and longer-term professionals have saturated the rental and owner-occupied units. The lack of affordable housing options affects the rural poor, children at risk, and homeless more than other populations. For example, a Bradford County key informant in Brasier et al.'s study stated: "For our homeless programs we would put people up at the the local hotel and we wanted to put someone up two weeks ago and the next available room is [four months later]. So there is no short-term housing."[60]

Quantitative studies of changes in housing prices offer mixed results. Farren et al. found that the number of wells drilled in a county is associated with increased fair market rent in Pennsylvania.[61] However, they also found that for the 144 counties studied in Pennsylvania, Ohio, New York, and West Virginia, there was no relationship between the number of wells drilled and median home value or vacancy rates. Similarly, a forthcoming study by McLaughlin and colleagues based on housing data from the U.S. Census and the American Community Survey find no consistent effects of Marcellus Shale development on housing values, vacancy rents, and housing affordability at the county level.[62] This is contrary to the findings from Kelsey and colleagues, who found that changes in market values between 2007 and 2009 were related to increased Marcellus activity in townships and boroughs, which suggests a more localized effect on housing markets.[63]

Health Services

There is little research documenting the effects of Marcellus Shale drilling activity on physical or mental health or on health care services.[64,65] In one of the few studies that have been conducted, researchers and clinical staff at the Southwest Pennsylvania Environmental Health Project found that the most common symptoms associated with drilling were skin rash or irritation, nausea or vomiting, abdominal pain, breathing difficulties or cough, and nosebleeds.[66] Another study by Ferrar and colleagues found that participants attributed 59 unique health impacts and 13 stressors to Marcellus Shale development.[67] In their quantitative analyses of health status and health care utilization data in four counties experiencing high levels of Marcellus Shale development, Davis and colleagues found no consistent changes related to health care access, hospitalization rates, and insurance rates; they did find increased levels of injuries associated with falls and motor vehicle accidents and emergency medical services complaints in the four counties.[68] Qualitative data in this same study suggest mixed impacts—significant stress on health care systems because of the increase in workers, the need to adjust to new insurance and health record systems, greater use of emergency medical services, and increased demand for substance abuse treatment—but also greater financial health because of an increase in employer-based health insurance.[69]

When discussing the health effects of development, workers' health is often ignored, yet workers who exhibit physical or mental health effects often strain local services.[70] Oil and gas development is a dangerous job, and many companies pride themselves on their health and environmental safety record, which helps the company to maximize profits and improve its public image. Nevertheless, there is a need for well-designed occupational health studies that assess the health rates of workers and the potential impacts they pose to the health care delivery system.

Local Government

Rapid population growth is expected to increase demand for services provided by local governments, including human services, physical infrastructure (roads, water, sewer), and land use planning. Often, local governments need to provide these services with few additional resources, or resources that arrive after the onset of demand. Impact studies conducted in Pennsylvania's Marcellus region show mixed or very few changes in revenues, expenditures, or staffing. A survey of local government officials across Pennsylvania's Marcellus region found that the majority (75%) of municipal governments said that the development activity had not affected their tax or nontax revenues, whereas about 18% said revenues had increased. About two-thirds reported that the public services they provided had not changed as a result of the activity.[71] Another study, focused on municipalities within Susquehanna and Washington (PA) counties found no clear impacts on municipal spending and revenues.[72] The passage of an impact fee (Act 13) in 2012 required energy companies to pay a per-well fee, a portion of which is to be provided to municipalities to compensate for expenses related to Marcellus shale development. Systematic analyses of the impacts of these funds and their use within Pennsylvania's municipalities and counties have not yet been completed.

Education

School districts are unique local institutions that could be substantially affected by rapid population changes. However, Schafft et al. found that no school districts in the affected counties with active drilling in Pennsylvania experienced a net increase in enrollment associated with Marcellus Shale development.[73] The researchers attribute the lack of increases, in contradiction to expectations, to a decrease in workforce demand due to technological innovations in the industry, an itinerant workforce that tends not to bring their families, and a lack of geographic isolation of Marcellus communities. Schafft et al. describe the views of school district personnel, who noted both opportunities for needed economic growth but also risks for environmental degradation and for uneven distribution of costs and benefits in Marcellus communities.[74]

Crime

Boomtown literature suggests that crime could increase because of several factors, including an overall increase in population, an increase in young males in the community, greater wealth in the community creating more opportunities for criminal acts, changes in reporting behavior, and the nature of the workers or the industry that increases the likelihood of criminal activity.[75,76] To date, the research documenting changes in crime associated with Marcellus Shale have shown mixed results. Kowalski and Zajac report no trends associated with Marcellus Shale development in calls-for-service data from the Pennsylvania State Police or arrest data from the Uniform Crime Reporting program through the Federal Bureau of Investigation.[77] The advocacy group Food and Water Watch examined crime rates for Pennsylvania counties and found that incidents of disorderly conduct increased in rural counties experiencing Marcellus Shale development at a higher rate than non-Marcellus counties.[78] In Brasier et al.'s examination of multiple indicators of criminal activity from 2001 through 2010, the researchers found that in counties experiencing the highest levels of drilling there were slightly higher reports of calls for service to which Pennsylvania State Police responded; slightly higher rates of serious crimes; and no substantial differences in arrest rates (serious, minor, DUI, drug abuse violations), criminal, civil, or traffic case filing rates; rates of sentences for misdemeanors; and county jail populations.[79]

Community Relationships and Conflict

A prominent area of research is how Marcellus Shale development has influenced residents' perceptions of their communities and the social relationships among community members. A consistent concern raised by study participants has been how the costs and benefits of development will be distributed among residents in the region, with a particular concern about how development may polarize the "haves" and the "have-nots."[80] Community leaders in the early stages of Marcellus Shale also described a desire to balance the economic benefits with concerns about how their community was changing, especially in relation to fundamental changes to the social relationships and physical beauty of the places that residents call home. Similarly, Jacquet and Stedman found that a boost to the local economy can create inequality, strain public services, increase divisions among community members, and alter one's sense of community.[81]

These community changes may instigate conflicts between community residents and newcomers to the community. Residents in Bradford County, Pennsylvania, for example, described the influx of energy companies and "foreign" energy workers as an "invasion" that negatively affected their way of life.[82] A quarter of survey respondents in one study reported that "the influx of new people had a 'substantial' or 'major' effect on the local area, in contrast with only 2.9% of respondents in lower drilling intensity areas."[83] These perceptions may manifest as real conflicts between long-term residents and newcomers. A participant in another study alluded to this conflict: "a lot of the workers who are coming here from other places have no ownership and therefore they don't feel the need to take care of this area. . . . They don't care if they trash the place or spend all their money on booze."[84] The extent to which controversial media portrayals of the gas industry and energy workers (e.g., Gasland) affect residents' perceptions is unknown; however, Filteau documents how itinerant energy workers perceive that long-term community residents stigmatize them as "dirty" and devalue their personal worth.[85] These workers also stigmatize other companies and workers that they perceive as unsafe as "dirtier."

Resident Risk Perception and Community Conflict

An important driver of community conflict is differing views on the risks and the opportunities—and who bears those risks and receives those opportunities—associated with development. Differences in perceptions of the risks are driven at least partially by the ability to directly benefit from development. Two separate analyses from the same representative mail survey of New York and Pennsylvania residents living in the Marcellus Shale region show how Pennsylvania participants who convey distrust in the natural gas industry perceive higher risks, whereas New York residents report higher opposition levels to development.[86] In addition to place of residence, leasing status also affects respondents' support and opposition for unconventional energy development. Kriesky et al. found that residents who leased their land were more supportive of energy development than residents without a lease.[87]

Contemporary Energy Boomtowns: Challenging and Extending the Boomtown Model

This wave of research on contemporary energy development has raised questions about how scholars should conceptualize booms and study the effects of energy extraction. Here we describe four questions raised by this research that challenge us to modify or develop a new model to understand the community impacts of energy development. This model emphasizes how the spatial, temporal, and cultural context of development shapes local impacts. We also highlight the need to understand the industry and regulatory drivers across the region. Several methodological issues also need to be addressed if we are to further our knowledge in this area. Previous boomtown studies did not capture individual and collective efforts for and against energy development; we highlight some of these efforts in the Marcellus Shale region. We conclude by identifying a number of research directions and theoretical traditions that enable us to advance beyond the existing research.

How do the spatial, temporal, and cultural contexts shape community impacts? Jacquet and Kaye identify a series of assumptions about the boomtown model that may not fit contemporary development contexts.[88] Two of these—level of isolation and the spatial concentration of development activity—are driven by the context of the communities in which the extraction is occurring. Isolated communities, such as those in the intermountain West where the boomtown model was developed, were assumed to have fewer influences on them, and as a result, the influx of natural resource development could be identified as the main cause of any social and economic changes. Communities in the Marcellus region are not as small or as isolated as those in the intermountain West. They have higher overall levels of economic activity and are more integrated into a broader regional context. This makes detecting changes associated with the introduction of one new economic activity, unless it is quite large, difficult. In communities that are less remote and more integrated regionally, to what extent can the effects of one economic activity be isolated from long-standing trends (such as population decline) or changes in the community or broader economy (such as the recession)?[89]

As Jacquet and Kaye note, early boomtown research conceived of the development as a single facility, circumscribed in time and space.[90] Contemporary shale-based development, however, is spread over a multicounty or even multistate region, and it is projected to last 40 to 50 years. The industry has so far shown a tendency to move drilling equipment and workers within and among "plays" in response to price changes and leasing arrangements. The result is a series of "mini-booms" and "mini-busts" within localities among the regions. Consequently, researchers studying boomtown effects need to take these localized stages of development into account when assessing community changes.

Furthermore, because of the spatial distribution of the natural gas resources, production companies and subcontractors have developed a system of regional headquarters and field offices to service the gas field. The headquarters tend to be centrally located to provide key infrastructure (worker housing, commercial space, transportation networks). Workers and subcontractors commute from those central locations into the region where the wells are located, creating complex, hub-and-spoke patterns, with the hubs varying by the stage of the process and the location. This pattern of development means that the boomtown effects—the population growth, service demand, social problems—tend to be concentrated within the "hub" cities and larger towns, with some diffusion across the surrounding region. The degree of this diffusion likely depends on the density of the transportation network that allows workers to move quickly around the region. The concentration of the effects may be greater in regions with a substantial disparity in population density and housing options between the central hub and the surrounding territory because the workers are more likely to be concentrated in the hubs in a region where there are few other options for services.

The nature of the industry itself—and the geological and ecological characteristics that influence the quality, accessibility, and economic potential of the energy resource—also need to be considered. For example, Luthra and colleagues argue that boomtown models simply do not apply to the offshore oil industry.[91] This is partly due to the remote nature of the activity but also to the historical, cultural, ecological, and economic relationships that develop over time, creating differential reactions to energy development.

The opportunity that the burgeoning oil and gas development offers is to develop a better understanding of how the local context shapes the impacts of this development on communities. Although there is research comparing communities within the same play, comparing energy booms across plays in the United States—for example, Marcellus, Barnett, Haynesville, Fayetteville, Bakken—may provide insight to the ways that context shapes community outcomes. Each region has differences in productivity, hydrocarbon content, and production techniques, as well as social, ecological, cultural, regulatory, and economic differences. As unconventional oil and gas development grows across the globe, there is growing interest in international comparisons to describe how the community impacts are shaped by property rights regimes and political systems.

How do industry and regulatory structures shape community impacts? Studies of community impacts of Marcellus Shale development examine the outcomes shaped by private business decisions, public and private organizational structures, and regulatory systems at local, state, and federal levels. For example, production companies, and their myriad subcontractors, tend to have particular regions in which they work based on their lease holdings. This concentration of a small set of production companies, and their subcontractors, can create a "footprint" on that area. This "footprint" is defined by the contractual and health and safety policies that shape the employment conditions of energy workers and the companies' community relations policies. These policies influence workers' behaviors in communities, behaviors that can have significant consequences related to housing, crime, health, and drug and alcohol abuse. We need to understand differences in companies' policies and organizational structures and how they shape communities.

What are effective research methodologies to document community impacts? There are a number of methodological issues associated with studying community impacts of boomtowns. Longitudinal and comparative research is needed to understand the impacts over phases of development. In the present era, this research is particularly important to understand the effects from successive stages of "mini booms" and "mini busts." There are additional challenges associated with the nature of contemporary development. As noted earlier, the hub-and-spoke pattern of development does not align with municipal or county boundaries, making it difficult to assess community impacts using traditional secondary data sources that rely on these geographic units. Counties as the unit of analysis are too large because the aggregate data are missing potential localized impacts of development.

In many Pennsylvania counties, for example, Marcellus Shale development is concentrated within a few townships rather than evenly distributed throughout counties. County-level analyses are likely to miss localized effects of development that could occur due to a higher concentration of activity in these areas. Possible changes in these localities may be hidden in the county-level data because they are unavoidably combined with information from the counties' less active areas. Regions may more appropriately capture the pattern of development of this industry but are not easily defined. Furthermore, the availability of data that would provide relevant longitudinal data for either smaller units of analysis or entire regions is limited. More research is needed to appropriately define hub-and-spoke patterns and the interactions between workers and their host communities.

An additional problem relates to the ways in which we identify the location and extent of oil and gas activity. Most studies of the impacts of Marcellus Shale development rely on well locations. Yet wells are only one element in a series of activities occurring both near and far from the wells, including pipeline construction, compressor stations, water withdrawal and storage sites, pipe and other storage yards, maintenance facilities, regional offices, worker housing, truck traffic, and road repairs.[92] It is difficult to identify and examine the impacts of an activity when a full accounting of those activities on the landscape are not known. For example, the natural gas boom has created a demand for sand used in hydraulic fracturing. The sand primarily comes from Wisconsin and Iowa, which has led to local community issues such as truck traffic, economic growth, demand for transportation, and environmental concerns related to sand mining.

So far, analyses of Marcellus Shale development have relied on wells as the primary metric for Marcellus Shale development, even though it incompletely represents the scope, breadth, and geographic locations of activities related to unconventional gas development activity. Additional research is needed that documents the networks of relationships and their impacts on local communities and environments at each stage of the energy development process. Such work could use as a foundation recent work in environmental flows that documents networks of relationships created by the flow of a commodity from extraction through to consumption.

What are effective public engagement strategies related to shale development? Shale-based development in the U.S. context is initiated through a private transaction (i.e., the execution of a lease between the subsurface rights owner and the production company), governed by relevant state and federal rules, and occurs with relatively little public input. Those opposed to the activity primarily advocate for changes in rule-making and regulatory procedures using social movement tactics such as protests, lobbying, and direct advocacy campaigns. However, in Pennsylvania, there have been few opportunities to influence the likelihood of extraction or the procedures used. Political engagement at the local level, such as involvement with local advisory bodies (e.g., energy task forces or committees), is one opportunity, but this does not foster broad public debate. Deliberative, dialogue-based approaches, such as public issues forums or citizen juries, may provide opportunities for learning and engagement but offer few opportunities to influence regulations.[93]

Perhaps partially related to the lack of public engagement opportunities, the emergence of local opposition has created substantial conflict in some communities, leading to concerns about the development of "corrosive communities."[94] Some researchers who frame responses in terms of feelings of inevitability suggest passive responses to the certainty of energy exploration and extraction.[95] However, there is also growing interest in the development of local opposition to shale development. Studies that draw on social movements literature might examine how and under what circumstances mobilization has occurred. Research that uses innovative social movement perspectives in concert with boomtown studies will better understand the "fracktivism" protest movements, passage of municipal antifracking bans, and growth of pro-development advocacy groups.

A critical element of conflict about shale-based energy development relates to the ways in which the potential risks are framed. Studies have examined differences among individuals in terms of their perceptions of the risks of shale gas development and specifically to the technology of hydraulic fracturing.[96,97] Findings suggest that perceptions of risks are directly influenced by trust in the natural gas industry. There are opportunities to further this research through multiple avenues, such as drawing on the social amplification of risk framework to describe, for example, the ways in which risk perceptions are influenced by sources of information (called "amplification stations" in this literature) and important symbolic events (e.g., spills, accidents, publication of major scientific findings). There is a need for additional research on the pathways through which individuals receive information about shale-based energy development and how they incorporate that information into their own perceptions of the risks.[98] Given the polarized nature of discussion about this issue in some communities, there is a need for a greater understanding of how risk perceptions are developed and influenced to enable more productive dialogue.

CONCLUSION

Although the boomtown model presents some limitations for studying contemporary energy booms, some of the theoretical and methodological insights to emerge from scholarly debates in the 1980s will benefit current and future inquiries. First, longitudinal research enables researchers to chronicle how contemporary energy development evolves with time, which will enable researchers to understand whether, and the extent to which, communities endure "mini booms" and "mini busts." Second, researchers should continue to study how energy development affects individuals and communities differentially. These studies will be strongest when they combine both quantitative and qualitative research, particularly given the limitations of secondary data and the importance of perceptions of change for subsequent individual and community decision-making. Studies also need to consider how, and to what extent, individuals and communities are able to influence outcomes through activism, advocacy, and regulatory change. Finally, we urge researchers to consider engaging with theoretical traditions that extend beyond the boomtown model to provide additional insight into growth of opposition and conflict and to identify effective mechanisms for public engagement.

NOTES

1. Pearson I, Zeniewski P, Gracceva et al. *Unconventional Gas: Potential Energy Market Impacts*. Petten, The Netherlands: European Commission Joint Research Centre, 2012.
2. Waples DA. *The Natural Gas Industry in Appalachia*. North Carolina: McFarland & Co., 2012.
3. Wilber T. *Under the Surface: Fracking, Fortunes, and the Fate of the Marcellus Shale*. Ithaca, NY: Cornell University Press. 2012.
4. U.S. Energy Information Administration. *Annual Energy Outlook*. http://www.eia.gov/forecasts/aeo/MTnaturalgas.cfm#natgas_prices?src=Natural-b1 (accessed May 2, 2014).
5. U.S. Energy Information Administration. *Energy in Brief*. http://www.eia.gov/energy_in_brief/article/major_energy_sources_and_users.cfm (accessed May 2, 2014).
6. Caulton DR, Shepson PB, Santoro RL, et al. Toward a better understanding and quantification of methane emissions from shale gas development. *Proceedings of the National Academy of Sciences* 111 (2014): 6237–6242.

7. Rahm BG, Bates JT, Bertoia LR, et al. Wastewater management and Marcellus Shale gas development: trends, drivers, and planning implications. *Journal of Environmental Management* 120 (2013): 105–113.

8. Coleman JL, Milici RC, Cook TA, et al. *Assessment of Undiscovered Oil and Gas Resources of the Devonian Marcellus Shale of the Appalachian Basin Province, 2011* (Fact Sheet 2011-3092). Reston, VA: U.S. Geologic Survey. 2011.

9. Engelder T. Marcellus. *Fort Worth Basin Oil & Gas Magazine* (August 2009): 18–22.

10. Milici RC, Swezey CS. *Assessment of Appalachian Basin Oil and Gas Resources: Devonian Shale—Middle and Upper Paleozoic Total Petroleum System* (Open-File Report Series 2006-1237). Reston, VA: U.S. Geologic Survey, 2006.

11. Pennsylvania Department of Environmental Protection Office of Oil and Gas Management. *Wells Drilled by County*. Harrisburg: Pennsylvania Department of Environmental Protection. http://www.portal.state.pa.us/portal/server.pt/community/oilandgasreports/20297 (accessed May 21, 2014).

12. U.S. Energy Information Administration. *Today in Energy*. http://www.eia.gov/todayinenergy/detail.cfm?id=14231 (accessed February 5, 2014).

13. US Energy Information Administration. *Annual Energy Outlook 2014*. http://www.eia.gov/forecasts/aeo/MT_naturalgas.cfm#natgas_gasfired (accessed February 5, 2014).

14. Freudenburg WR, Wilson LJ. Mining the data: Analyzing the economic implications of mining for nonmetropolitan regions. *Sociological Inquiry* 72, no. 4 (2002): 549–575.

15. White NE. A tale of two shale plays. *The Review of Regional Studies* 42, no. 2 (2013): 107–119.

16. England JL, Albrecht SL. Boomtowns and social disruption. *Rural Sociology* 49 (1984): 230–46 (p. 231).

17. Gilmore JS, Duff MK. *Boom Town Growth Management: A Case Study of Rock Springs-Green River, Wyoming*. Boulder, CO: Westview Press, 1975.

18. Cortese CF. The social impacts of energy development in the west: An introduction. *The Social Science Journal* 16 (1979): 2.

19. Gilmore JS. Boom Towns may hinder energy resource development: Isolated rural communities cannot handle sudden industrialization and growth without help. *Science* 191 (1976): 535–540.

20. Kohrs EV. Social consequences of boom growth in Wyoming. Paper presented at the Rocky Mountain American Association of the Advancement of Science Meeting, April 24–26, 1974; Laramie, Wyoming. http://www.sublettewyo.com/archives/42/Social_Consequences_of_Boom_Growth_In_Wyoming_-_Kohrs[1].pdf (accessed November 12, 2014).

21. Lantz AE, McKeown RL. *Social/psychological problems of women and their families associated with rapid growth*. U.S. Commission on Civil Rights Energy Resource Development. Washington, DC: U.S. Government Printing Office, 1979.

22. Freudenburg WR, Jones RE. Criminal behavior and rapid community growth: examining the evidence. *Rural Sociology* 54, no. 4 (1991): 619–645.

23. Freudenburg W. The impacts of growth on the social and personal well-being of local community residents. In B. Weber and R. Howell, eds. *Coping with Rapid Growth in Rural Communities*, pp. 137–170. Boulder, CO: Westview Press, 1982.

24. Brown BS. *The Impact of the New Boomtowns: The Lessons of Gillette and the Powder River Basin*. Rockville, MD: U.S. Dept. of Health, Education, and Welfare, Public Health Service, Alcohol, Drug Abuse, and Mental Health Administration, National Institute of Mental Health, 1978.

25. Gilmore JS, Duff MK. *Boom town growth management: A case study of Rock Springs-Green River, Wyoming*. Boulder, CO: Westview Press, 1975.

26. Freudenburg, The impacts of growth, op. cit.

27. Leistritz FL, Murdock SH, Leholm AG, et al. Local economic changes associated with rapid growth. In B. Weber and R. Howell, eds. *Coping with Rapid Growth in Rural Communities*, pp. 25–61. Boulder, CO: Westview Press 1982.

28. Lovejoy SB, Little RL. Energy development and local employment. *The Social Science Journal* 16, no. 2 (1979): 169–190.

29. Wilkinson KP, Thompson JG, Reynolds RR Jr, et al. Local social disruption and western energy development: a critical review. *Pacific Sociological Review* 25, no. 3 (1982): 275–296.

30. Ibid.

31. Freudenburg W. An overview of the social science research. In CM McKell, DG Browne, EC Cruze, et al., eds. *Paradoxes of Western Energy Development: How Can We Maintain the Land and the People If We Develop?* Boulder, CO: Westview Press, 1984.

32. Ibid.

33. Wilkinson et al., Local social disruption, op. cit.

34. Brown R, Dorius S, Krannich R. The boom–bust–recovery cycle: dynamics of change in community satisfaction and social integration in Delta, Utah. *Rural Sociology* 70 (2005): 28–49.

35. Smith MD, Krannich RS, Hunter LR. Growth, decline, stability, and disruption: a longitudinal analysis of social well-being in four western rural communities. *Rural Sociology* 66 (2001): 425–450.

36. Brown et al., The boom–bust–recovery cycle, op. cit.

37. Hunter LM, Krannich RS, Smith MD. Rural migration, rapid growth, and fear of crime. *Rural Sociology* 67 (2002): 71–89.

38. Smith et al., Growth, decline, stability, and disruption, op. cit.

39. Brown R, Geertsen HR, Krannich R. Community satisfaction and social integration in a boom-town: a longitudinal analysis. *Rural Sociology* 54 (1989): 568–586.

40. Brown et al., The boom–bust–recovery cycle, op. cit.

41. Smith et al., Growth, decline, stability, and disruption, op. cit.

42. Forsyth J, Luthra AD, Bankston WB. Framing perceptions of oil development and social disruption. *The Social Science Journal* 44 (2007): 287–299.

43. Brown TC, Bankston WB, Forsyth CJ, Berthelot ER. Qualifying the boom-bust paradigm: an examination of the off-shore oil and gas industry. *Sociology Mind* 1 (2011): 96–101.

44. Luthra AD, Bankston WB, Kalich DM, et al. Economic fluctuation and crime: a time-series analysis of the effects of oil development in the coastal regions of Louisiana. *Deviant Behavior* 28, no. 2 (2007): 113–130.

45. Freudenburg WR. Addictive economies: extractive industries and vulnerable localities in a changing world economy. *Rural Sociology* 57 (1992): 305–332.

46. Jacquet J. *Energy Boomtowns and Natural Gas: Implications for Marcellus Shale Local Governments and Rural Communities* (Rural Development Paper, No. 43). State

College, PA: Northeast Regional Center for Rural Development, 2009. http://nercrd.psu.edu/Publications/rdppapers/rdp43.pdf (accessed May 20, 2014).

47. Freudenburg, Addictive economies, op. cit. (p. 306).

48. Brown et al., The boom–bust–recovery cycle, op. cit.

49. Considine TJ, Watson R, Blumsack S. *The Economic Impacts of the Pennsylvania Marcellus Shale Natural Gas Play: Status, Economic Impacts and Future Potential.* Philadelphia: The Pennsylvania State University, 2010.

50. Kinnaman TC. The economic impact of shale gas extraction: a review of existing studies. *Ecological Economics* 70 (2011): 1243–1249.

51. Herzenberg S. *Drilling Deeper into Job Claims: The Actual Contribution of Marcellus Shale to Pennsylvania Job Growth.* Harrisburg, PA: Keystone Research Center, 2011.

52. Kelsey T, Shields M, Ladlee JR, Ward M. *Economic Impacts of Marcellus Shale in Pennsylvania: Employment and Income in 2009.* 2011. http://www.shaletec.org/docs/economicimpactfinalaugust28.pdf (accessed November 12, 2014).

53. Kelsey TW, Hardy K, Glenna L, et al. *Local economic impacts related to Marcellus Shale development*. Report prepared for the Center for Rural Pennsylvania. http://www.rural.palegislature.us (accessed November 27, 2014).

54. Kelsey et al., Local economic impacts related to Marcellus Shale development, op. cit.

55. Jacobson M, Kelsey TW. Impacts of Marcellus Shale development on municipal governments in Susquehanna and Washington Counties, 2010. Penn State Extension Marcellus Education Fact Sheet. http://www.marcellus.psu.edu/resources/pdfs/Jacobson_fiscal.pdf (accessed November 27, 2014).

56. Kelsey et al., *Economic Impacts of Marcellus Shale*, op. cit.

57. McLaughlin D, Rhubart D, DeLessio-Parson A, et al. *Population Change and Marcellus Shale Development*. Report prepared for the Center for Rural Pennsylvania. http://www.rural.palegislature.us (accessed November 27, 2014).

58. Williamson J, Kolb B. *Marcellus Natural Gas Development's Effect on Housing in Pennsylvania.* Williamsport, PA: Lycoming College Center for the Study of Community and the Economy. http://www.housingalliancepa.org/sites/default/files/resources/Lycoming-PHFA%20Marcellus_report.pdf (accessed May 20, 2014).

59. Ooms T, Tracewski S, Wassel K, et al. *Impact on Housing in Appalachian Pennsylvania as a Result of Marcellus Shale*. Wilkes-Barre, PA: The Institute for Public Policy and Economic Development,. http://www.institutepa.org/PDF/Marcellus/housing11.pdf (accessed May 20, 2014).

60. Brasier KJ, Filteau MR, McLaughlin DK, et al. Residents' perceptions of community and environmental impacts from development of natural gas in the Marcellus Shale: a comparison of Pennsylvania and New York cases. *Journal of Rural Social Sciences* 26 (2011): 32–61 (p. 52).

61. Farren M, Weinstein A, Partridge M, et al. *Too Many Heads and Not Enough Beds: Will Shale Development Cause a Housing Shortage?* Columbus: The Swank Program in Rural-Urban Policy, The Ohio State University. http://go.osu.edu/shale_housing_rpt (accessed May 20, 2014).

62. McLaughlin D, DeLessio-Parson A, Rhubart D. *Housing and Marcellus Shale Development.* The Marcellus Impacts Project Report #5. www.rural.palegislature.us/.../reports/Marcellus-Report-5-Housing.pdf (accessed November 27, 2014).

63. Kelsey TW, Adams R, Milchak S. *Real Property Tax Base, Market Values, and Marcellus Shale: 2007–2009* (CECD Research Paper Series). 2012. http://cecd.aers.psu.edu (accessed May 20, 2014).

64. Goldstein BD, Bjerke EF, Kriesky J. Challenges of unconventional shale gas development: so what's the rush? *Journal of Legal Ethics and Public Policy* 27 (2013): 149–186.

65. McDermott-Levy R, Kaktins N. Preserving health in the Marcellus region. *Pennsylvania Nurse* 67, no. 3 (2012): 4–10.

66. Kreisky J. *Health Issues and Concerns Related to Unconventional Gas Development*. Southwest Pennsylvania Environmental Health Project. http://www.environmentalhealthproject.org (accessed May 20, 2014).

67. Ferrar KJ, Kriesky J, Christen CL, et al. Assessment and longitudinal analysis of health impacts and stressors perceived to result from unconventional shale gas development in the Marcellus Shale region. *International Journal of Occupational and Environmental Health* 19 (2013): 104–112.

68. Davis LA, McLaughlin D, Uberoi N. *The Impact of Marcellus Shale Development on Health and Health Care*. Report prepared for the Center for Rural Pennsylvania.http://www.rural.palegislature.us/documents/reports/Marcellus-Report2-Health.pdf (accessed November 27, 2014).

69. Ibid.

70. Goldstein et al., Challenges of unconventional shale gas development, op. cit.

71. Kelsey TW, Ward MM. *Natural Gas Drilling Effects on Municipal Governments throughout Pennsylvania's Marcellus Shale Region, 2010* (Penn State Cooperative Extension Marcellus Education Fact Sheet). University Park, PA: Penn State Cooperative Extension, 2011.

72. Jacobson and Kelsey, *Impacts of Marcellus Shale Development*, Op. cit.

73. Schafft KA, Glenna LL, Green B, et al. Local impacts of unconventional gas development within Pennsylvania's Marcellus Shale Region: gauging boomtown development through the perspectives of educational administrators. *Society & Natural Resources* 27 (2013): 389–404.

74. Schafft KA, Kotok S, Biddle C. *Marcellus Shale Gas Development and Impacts on Pennsylvania Schools and Education*. Report prepared for the Center for Rural Pennsylvania. http://rural.palegislature.us/reports/The-Marcellus-Shale-Impacts-Study (accessed November 27, 2014).

75. Ruddell R. Boomtown policing: responding to the dark side of resource development. *Policing* 5 (2011): 328–342.

76. Ruddell R, Jayasundara DS, Mayzer R, Heitkamp T. Drilling down: an examination of the boom–crime relationship in resource-based boom counties. *Western Criminology Review* 15 (2014): 3–17.

77. Kowalski L, Zajac G. *A Preliminary Examination of Marcellus Shale Drilling Activity and Crime Trends in Pennsylvania*. University Park: Justice Center for Research, Pennsylvania State University. http://justicecenter.psu.edu/research/documents/MarcellusFinalReport.pdf (accessed May 23, 2014).

78. Food and Water Watch. *The Social Costs of Fracking: A Pennsylvania Case Study*. http://documents.food andwaterwatch.org/doc/Social_Costs_of_Fracking.pdf (accessed May 22, 2014).

79. Brasier K, Rhubart D. *Effects of Marcellus Shale Development on the Criminal Justice System. Report prepared for the Center for Rural Pennsylvania,*

Harrisburg, Pennsylvania. http://rural.palegislature.us/reports/The-Marcellus-Shale-Impacts-Study (accessed November 27, 2014).

80. Brasier et al., Residents' perceptions of community and environmental impacts, op. cit.

81. Jacquet JB, Stedman RC. The risk of social-psychological disruption as an impact of energy development and environmental change. *Journal of Environmental Planning and Management* 57 (2013): 1–20.

82. Perry S. Development, land use, and collective trauma: the Marcellus Shale gas boom in rural Pennsylvania. *Culture, Agriculture, Food and Environment* 34 (2012): 81–92.

83. Schafft et al., Local impacts of unconventional gas development, op. cit. (p. 6).

84. Brasier et al., Residents' perceptions of community and environmental impacts, op. cit. (p. 47).

85. Filteau, MR. Who are those guys? Constructing the oilfield's new dominant masculinity. *Men and Masculinities* 17 (2014): 396–416.

86. Stedman RC, Jacquet JB, Filteau MR, et al. Environmental reviews and case studies: Marcellus Shale gas development and new boomtown research: views of New York and Pennsylvania residents. *Environmental Practice* 14 (2012): 382–393.

87. Kriesky J, Goldstein BD, Zell K, et al. Differing opinions about natural gas drilling in two adjacent counties with different levels of drilling activity. *Energy Policy* 58 (2013): 228–236.

88. Jacquet JB, Kay DL. The unconventional boomtown: updating the impact model to fit new spatial and temporal scales. *Journal of Rural Community Development* 9 (2014): 1–23.

89. McLaughlin et al., *Population Change and Marcellus Shale Development*, op cit.

90. Jacquet and Kay, The unconventional boomtown, op cit.

91. Luthra et al., Economic fluctuation and crime, op cit.

92. Brasier K, Davis L, Glenna L, et al. *The Marcellus Shale Impacts Study: Chronicling Social and Economic Change in the Northern Tier and Southwest Pennsylvania.* Report prepared for the Center for Rural Pennsylvania. http://rural.palegislature.us/reports/The-Marcellus-Shale-Impacts-Study. (accessed November 27, 2014).

93. Centre County Public Issues Forum. *The Marcellus Shale: What Does It Mean for Us?* 2012. http://www.ontheridgeline.com/downloads/Marcellus%20draft.FINAL.pdf (accessed November 12, 2014).

94. Jacquet J, Stedman RC. Using risk analysis to measure social-psychological disruption as an impact of energy development and environmental change. *Journal of Environmental Planning and Management* 57 (2013): 1285–1304.

95. Malin S. There's no real choice but to sign: neoliberalization and normalization of hydraulic fracturing on Pennsylvania farmland. *Journal of Environmental Studies and Sciences* 4 (2014): 17–27.

96. Caulton et al., Toward a better understanding, op. cit.

97. Brasier et al., Risk perceptions of natural gas development, op. cit.

98. Boudet H, Clarke D, Bugden D, et al. "Fracking" controversy and communication: using national survey data to understand public perceptions of hydraulic fracturing. *Energy Policy* 65 (2014): 57–67.

Chapter 8

Risks beyond the Well Pad: The Economic Footprint of Shale Gas Development in the United States

Susan Christopherson

High-volume hydraulic fracturing is perhaps the most important industrialization process to occur in the United States for decades. This technology, also known as "massive horizontal slickwater hydraulic fracturing" or unconventional gas extraction (UGE) was first used in the mid-1990s in the Barnett Shale play of northeast Texas. It emerged out of many years of experimentation with techniques to profitably obtain oil and gas from shale deposits deep under the surface.

Although attractive because of its potential contribution to the U.S. balance of payments and to displace coal in domestic energy production markets, the development of unconventional gas is something less than an unalloyed "good." Those who are advocates of this technology speak of the benefits of the United States becoming not only energy independent but also a major exporter of natural gas, with all the economic and political benefits that this would create. Opponents present cogent arguments pointing to the potentially negative implications for the environment and well-being of the places where shale gas development occurs. Too often, however, assessment of the impact of hydraulic fracturing is narrowly focused. The goal of this chapter is to step back to view hydraulic fracturing through a wider lens, considering its extensive impacts across the American economic landscape.

Unlike iron ore and coal mining, which are located in less-populated regions such as Appalachia or the Powder River Basin in Montana and Wyoming, shale gas and oil extraction has a broader national geographic footprint. Extraction is possible in many locations, including areas where extractive industries have existed for generations and others where extraction is a new phenomenon. Urban and suburban communities may be sites for hydraulic fracturing as well as the more stereotypical rural, less-populated locations. In addition, the processes associated with providing the inputs to hydraulic fracturing and disposing of the toxic materials produced in unconventional gas extraction affect many cities and regions not directly engaged in the extraction process. To fully understand the implications of high volume hydraulic fracturing on the U.S. economy, we need to look beyond the well pad. Specifically, what can previous experience tells us about the regional economic impacts of extraction-based resource development? What types of local economies are affected by the industrial processes connected to hydraulic fracturing? How are the differences among those places likely to alter the nature and extent of the impacts? How is hydraulic fracturing affecting regional economies where no shale gas or oil development is taking place?

Because unconventional gas extraction takes place in many types of environments, impacts will vary and be more "visible" in some places than others. One of the challenges in systematically identifying and accounting for economic impacts is taking into account this variability. For example, some impacts, such as increases in public safety costs related to an increase in crime, are more visible in isolated communities (e.g., Dickinson, North Dakota) that fit the traditional depiction of the "mining boom town." In urban locations, such as Denton, Texas, an increase in crime may be absorbed in broader regional metropolitan patterns and not be as visible. Other types of costs, such as traffic accidents and congestion, may be intensified in suburban or urban extraction locations but have less of an impact in rural areas.

The systematic analysis of economic impacts is hampered by a lack of baseline data from which to monitor change. With the exception of county crime statistics maintained by the Federal Bureau of Investigation under the Uniform Crime Reporting Statistics program, there are no data compilations that support comparison of social or economic impacts across counties. Because of different modes of reporting well permitting, production, and completion across states, it is not possible to analyze the pattern of economic impacts and public costs as it relates to the progress of the drilling cycle. Data definitively documenting how localities are affected by shale gas and oil development are currently not available because neither the states nor the federal government have been willing to collect it. Data, if available at all, must be assembled state-by-state, county-by-county, or agency-by-agency.

Even when data are collected, the lack of comparative statistics across states and localities makes a systematic analysis of the different types of impacts almost impossible. For example, a recent small comparative study of monitoring of public health complaints arising from UGE in three states found that Wyoming did not record any health complaints; North Dakota has started recording complaints but the data are not made public; Colorado, in contrast, both records complaints and makes them public.[1] As such, the absence of data on impacts has hampered states' ability not only to realistically assess the costs of UGE development but also to impose impact fees or taxes to compensate for potential losses.

Despite the absence of statistical data, a growing body of literature is documenting similar economic impacts among localities in different shale plays and in natural resource extraction economies. This literature includes environment impact statements, public policy reports, academic journal articles, and eyewitness accounts by journalists. In some cases, there are multiple accounts of economic impacts in the same area at different points in time—for example, Sublette County Wyoming, and Williston, North Dakota. This literature forms the evidence used in this chapter.

This chapter discusses what is known about the economic impact of resource extraction on regional economies with particular attention to what has been learned about UGE economies in U.S. shale plays. An examination of the different types of regions affected by UGE is presented, including how differences among those regions may affect the range, visibility, and intensity of economic impacts. Finally, I broaden the lens to look at how regions outside the UGE extraction areas are affected by the national industrialization process and what a wider perspective implies for policy.

WHAT WE KNOW ABOUT THE ECONOMIC IMPACTS OF RESOURCE-BASED EXTRACTION ECONOMIES—THE BOOM–BUST CYCLE

The United States has a rich lore of "boomtowns" and "ghost towns," yet people rarely connect this history—and the boom–bust cycle it depicts—to contemporary resource development.[2–4] Today's unconventional gas extraction using horizontal drilling and high volume hydraulic fracturing is both similar to and different from previous experience. UGE undoubtedly will produce a cycle of boom and bust at the local level. Like any nonrenewable resource development, shale gas development does bring an economic "boom" to extraction regions, at least during the period when drilling sites and support facilities are set up and drilling takes place. As drilling companies move into a community, population flows in for employment or to "cash in" on the boom. Local expenditures rise on everything from auto parts to pizza and beer. There also is an increase in jobs outside the extraction industry itself in construction, transportation, retail, hotels and restaurants, entertainment, and services.[5,6]

Landowners receive royalty payments and have extra money to spend. The tax base may expand, providing a windfall for a local government. However, research on actual employment impacts in resource development regions indicates that job projections are typically overstated.[7,8] For example, in the seven states in the United States with more than 5,000 employees in the oil and gas extraction industry (Texas, California, Oklahoma, Pennsylvania, Colorado, Louisiana, and New Mexico), the percent of state employment in the oil and gas industry is well under 1% of total state employment.[9] Although employment increased between 2002 and 2012 in these states, the *percentage* remained under 1% except in Oklahoma, where it totaled 1.5% of total state employment in 2012.[10] The "high growth rates" used to indicate the industry contribution to employment often ignore the reality that the industry employs small numbers of people.

Notwithstanding the exaggerated estimates that abound regarding job impacts, the increased economic activity associated with UGE is welcome in some communities, especially among individuals who are expecting or hoping to reap direct economic benefits. However, although a natural resource extraction boom may bring jobs and population growth for a few years, it also increases public service costs and the cost of living for residents, "crowds out" other industries, and may raise their cost of doing business.[11] In the case of UGE, these other industries may include tourism, retirement communities, manufacturing, or organic agriculture.

Shale gas development also brings an additional level of uncertainty to regional economic forecasting. Because a substantial number of U.S. states are engaged in shale gas and oil extraction, some producing dry gas and some with higher-profit oil deposits, drilling rigs may move at short notice from one region to another, causing a series of economic disruptions as drilling starts up, shuts down, and then starts up again.[12] This phenomenon has been affecting Pennsylvania since 2012. After several boom years, rigs and jobs have disappeared, leaving some areas with uncompleted wells and sharply reduced fees, as well as uncompleted building projects started during the boom period. Natural gas rigs in Pennsylvania dropped from a high of 112 operating in the state in 2011 to 51 in August 2013.[13]

This process of unpredictable boom and bust is inherent to resource extraction economies. Boomtowns frequently experience problems brought about by the influx of a transient population that follows the oil and gas industry rigs from one place to another. After the boom ends (either temporarily or permanently) and the drilling crews and their service providers depart, it is not inconceivable that the region may have a smaller population and a poorer economy than before the extraction industry moved in.[14] If the boom–bust cycle is combined with environmental damage, the long-term costs to regions hosting shale gas and oil extraction may be considerable because they will limit future investment and tourism.

What does all this mean? Essentially, natural resource development—including unconventional gas extraction—is positive for some segments of the population (mineral rights owners, some businesses) and negative for others (renters, landowners without mineral rights, businesses in competing industries). When the commercially viable resources are depleted, drilling ceases—either temporarily or permanently—and there is an economic "bust," as businesses and personnel connected to resource extraction leave the community.[15,16] Mineral rights owners may continue to derive royalties from their leases, but the impact of those royalties on the regional economy is unclear.[17] Mineral rights lease-holders may not reside in the region, may invest rather than spend their royalties, or may spend their royalties outside the region. So in addition to issues related to the boom–bust nature of economic development in UGE regions, there are complex questions about how the boom period economic benefits are distributed in the population and geographically.

LOOKING REGIONALLY: PRODUCTION SITES MAY NOT REAP ECONOMIC BENEFITS, AND COSTS MAY BE DISPLACED

Although jobs are created in drilling regions, evidence shows that most well site jobs go to outsiders (e.g., drilling contract companies).[18] Local residents get some jobs in support of drilling activity (e.g., in lodging, food and entertainment, or retail services), but most high-paying jobs go to those who are brought in to drill the wells.[19] Moreover, jobs may not be created in the communities where hydraulic fracturing is occurring. That is, they may be created in another county, or another state. This will become clearer in the subsequent pages.

The complex regional character of UGE on the economy is well illustrated by what is happening in the Marcellus Shale, specifically in northern Pennsylvania and the Southern Tier of New York. While drilling activity is confined to the northern Pennsylvania counties (because there is a moratorium on drilling in effect in New York State), many of the economic benefits associated with UGE accrue to the Southern Tier New York counties primarily because these counties have commercial facilities, including hotels, restaurants, and retail establishments in place. Sales tax receipts in the New York counties along its southern border with Pennsylvania increased at the highest rate in the state during the height of the drilling boom from 2010 through 2012.[20] This displacement of economic benefits from the neighborhood of the drilling sites to other locations in the regional economy, part of which is in another state, demonstrates an important principal of economic impacts related to resource extraction. It is not where the drilling activity occurs per se, but where the expenditures are made that determines the location of economic benefits from resource development-driven industrialization.

The complexity of predicting where benefits of UGE will occur also applies to the costs of natural gas development. UGE is a regional and national industrial process, and the costs of natural gas development may affect places far from the well sites. State and local governments—counties, cities, townships, villages—in states where UGE is taking place are coping with demands for new services and increased levels of service. The administrative capacity, staffing levels, equipment, and outside expertise needed to meet those demands may be beyond what has been budgeted. In Sublette County, Wyoming, for example, as the number of gas wells drilled *per year* climbed from 100 in 2000 to more than 500 in 2006, during this time period the population of Sublette County swelled by 24%, but Wyoming's population grew by just 4%, indicating that workers and their families were flocking to the area to meet the new labor demand. The most dramatic increase in population came from teens and young adults aged 15 to 24. As the age cohort 25 to 44 years was decreasing statewide, it was increasing in the county. Indeed, all cohorts of working age adults increased more rapidly in Sublette County than statewide.[21] This short-term population influx created significant demands on public services.[22]

Both the Sublette County experience and the now well-documented experience of North Dakota communities of Dickinson and Williston[23] indicate that UGE increases a wide range of public service costs, many paid for at the state level. In Sublette County, Jacquet[24] found that traffic on major roads increased, as did the number of traffic accidents, the number of emergency room visits, and the demand for emergency response services. In addition, local schools experienced increased demand, as some workers moved their families to the region and had to enroll their children in school. As demand for all manner of goods and services increased and local businesses sought to exploit the boom, prices went up. Jacquet found that local prices in Sublette County increased by twice the national rate over a 6-year "boom" period.

The price inflation characteristic of shale boom areas especially affects rental housing. The drilling boom period in Williston, North Dakota, brought an instant population influx similar to that in Sublette County, Wyoming, leading to a homeless rate above 20%.[25,26] Williston has had previous experience with the boom–bust cycle of oil and gas development, and that experience has discouraged investment in the housing. Developers have been slow to build more apartments, largely because they were stung by the region's last oil boom that went bust in the 1980s.[27] The available evidence indicates that this largely rural region is having difficulty maintaining public services and public safety in the face of boomtown conditions.[28] The costs of this boom, however, are not all paid in Williston. Much is displaced to the state.

A rapid increase in UGE activity is not always associated with a commensurate increase in *resident population* in the counties where the drilling occurs. An analysis of population change in core natural gas drilling counties in the Marcellus Shale during the first decade of the 2000s, for example, found that the resident population in these largely rural counties has grown marginally if at all.[29] There are various reasons resident population growth does not occur in the core counties; the most frequently cited are the absence of services, the higher cost of living, and the lower quality of life in an industrialized environment. A reporter interviewing drillers who resided in neighboring New York State but who worked in Pennsylvania captured the reason in one quote: "There is nothing there [in Pennsylvania]—there's no entertainment, there's nothing to do . . . Chemung County [in New York State] is where we spend our money."[30]

When the 2012 decline in drilling for gas occurred in Northern Tier counties of Pennsylvania, with rigs and crews leaving the area, it was the Southern Tier of New York State that experienced a loss in sales tax revenues and customers. Southern Tier New York counties went from having the fastest growing sales tax revenues in the State of New York to, in 2014, have the steepest declines in sales tax revenue.[31]

Thus, when we think about the impacts of UGE, we see that there is no natural congruence between UGE and the economic benefits or costs of gas development. Moreover, because UGE and its associated supply chain activities take place in so many different types of regions, analyzing and evaluating UGE impacts is quite complex. The next section describes some of this complexity and its implications.

REGIONAL IMPACTS ACROSS A COMPLEX FRACTURING LANDSCAPE

When one looks beyond the well pad to assess how hydraulic fracturing affects places—communities and regions—one may be surprised at where things happen but also at the geographic extent of UGE. Although media attention generally is focused on the well site and the immediate neighborhood of households or on the local municipalities and hamlets adjacent to the drilling sites, an extensive multicounty region can be affected by drilling activities.[32] For example, the farming and ranching region above the Bakken Shale extends across North Dakota and Montana as well as two Canadian provinces, Manitoba and Saskatchewan. This large region includes small cities that are within the shale play (Williston and Minot, North Dakota). There are other small cities (e.g., Dickinson, North Dakota) that are outside the shale play but which are strongly affected by the economic and industrial activities connected to oil extraction using unconventional extraction techniques in the Bakken play. Although well pads are located in specific areas, the process also includes staging areas, pipelines, compressor stations, storage facilities, rail trans-shipment sites as well as thousands of trucks hauling chemicals, water, and the contaminate waste produced by the drilling process. Rural roads previously used primarily by farmers now have 800 trucks traversing them in a single day.[33] Flowback and produced water from the wells has to be transported to treatment facilities, which must be equipped to handle the increased volume and particular array of toxic and nontoxic wastes, or to injection wells. The facilities required will be located where geologic or logistical factors dictate; but, as described in more detail later in the chapter, these operations may touch communities hundreds of miles from the drilling regions, often in another state.

UGE industrial facilities create a wide range of intersecting environmental, economic, and social stressors, all of which have implications for the regional economy and its existing industries.[34–36] For example, noise is a major byproduct of compressor stations, which produce noise levels in the 85- to 95-decibel range.[37] These levels more often are at or above the U.S. Occupational Safety and Health Administration threshold of safety for an 8-hour day, and compressors work a 24-hour day. Environmental stressors can have an effect on the nearby population, adjacent property values, and on other industries in the vicinity, including those in adjacent urban neighborhoods.

One example of the impact shale development facilities may have on an urban or rural UGE industrial region is illustrated by the proposed gas storage facility in the Finger Lakes region of New York State, a major area for tourism because of its scenic beauty, small towns, and commercial vineyards. This facility, designed to have underground storage to hold 1.45 billion cubic feet of natural gas, is being planned by Inergy Midstream, LLC at the former salt plant just north of Watkins Glen, New York. There are plans to add an underground liquid propane storage facility designed to hold almost 89 million gallons and two large brine ponds above ground. The site for this major facility is near the intersection of two gas transmission pipelines. But Watkins Glen, in largely rural Schuyler County, is not part of the "fairway"—the purported "sweet spot" for Marcellus drilling in New York—and is many miles from the extraction sites in Pennsylvania whose production it would store. This storage site would not benefit from taxes or have an impact on fees obtained from extraction sites in Pennsylvania, but it will have infrastructure impacts in New York and will require professionally trained public sector emergency personnel to be located in Watkins Glen.

Whatever the plant may contribute in the way of local property taxes, the Watkins Glen economy and its jobs currently depend on Finger Lakes tourism, attendance at its famous auto races, the local wine industry, and agriculture. Local residents and businesses are concerned that this storage facility poses environmental risks, including the possibility of explosions. There are also risks from leaks from the facility and brine pond seepage or overflow that may affect the quality of water in Seneca Lake, which provides drinking water to thousands of central New York residents. Finally, tourist businesses and wineries fear the negative reputational effects that this type of industrial facility will have on important local businesses, especially the wineries. For all these reasons, Watkins Glen is a center of opposition to UGE despite its location in a state where UGE is not currently permitted.[38]

To some extent, this opposition can be explained by the changing character of and residential patterns in "rural" counties. Although Schuyler county is rural, by conventional definitions, with a population of about 20,000, many residents commute to work in the more populous and urban neighboring Tompkins County located to the east of Schuyler. They have jobs unrelated to agriculture or extraction industries and have purchased homes in Schuyler County because of amenities such as its wineries and its scenic beauty. Many oppose the natural gas storage facility because they strongly believe that it will affect the quality of life that drew them to this rural county in the first place. This is a perfect example of NIMBY—not in my backyard.

THE IMAGINED AND REAL LANDSCAPE OF HYDRAULIC FRACTURING

When people think about UGE, they picture isolated places that are sparsely settled and far from cities and suburbs. However, rural counties (e.g., counties not adjacent to metropolitan area counties) currently compose only half of the U.S. counties where shale gas and oil extraction is taking place.[39] This leaves another half that are classified as metropolitan or micropolitan (between 10,000 and 49,999 residents), places that also experience the boomtown impacts of UGE development. It is important to understand that UGE is either planned for or is ongoing in at least 28 states in areas that are characterized by many types of economic and environmental regions, including rural and semirural residential neighborhoods and communities. As was described, the transport of natural gas and its byproducts traverses many rural and urban areas. As a consequence, it is fair to state that shale gas development is not limited to one specific small area.

The boomtown character of development in the energy extraction regions is aptly demonstrated by Census Bureau estimates of the fastest growing metropolitan and micropolitan areas in the United States. Of the 10 fastest growing metropolitan statistical areas, six were within or adjacent to oil and gas fields: Odessa, TX; Midland, TX; Fargo, ND; Bismarck, ND; Casper, WY; and Austin-Round Rock, TX.[40] Among micropolitan statistical areas, seven of the fastest growing were located in or near oil and gas plays, with Williston, ND, ranked first in growth, followed by Dickinson, ND, and Andrews, TX. Again, it needs to be recognized that these micro areas have very small populations in actual real numbers. For example, Williston's population rose from 14,716 to 18,532 in 2010–2012, with estimates of another 11% rise in 2012–2013.[41] Although this may make Williams County, North Dakota, one of the fastest growing towns in the United States, the actual population is a fraction of that in the vast majority of U.S. metropolitan counties.[42] As was described in the previous section, the boom–bust character of resource extraction suggests that the micropolitan communities will shrink in size once the drilling phase of extraction is completed.

Thus, when we try to understand the impacts of UGE on communities and regions, we need to consider the scale of development in the region *before* gas drilling and the capacity of the community to absorb the costs that accompany development. In cities, such as Fort Worth or Denton, Texas, where oil and gas extraction has been extensive, there were emergency services and public safety personnel in place before the drilling boom. There is also rental housing stock to absorb workers migrating into the region and full-time professional managers in city offices charged with governing oil and gas development. In micropolitan areas, however, city managers often work part-time, and emergency personnel may be made up of resident volunteers. On the other hand, some impacts—such as road congestion, noise from compressor stations, and citizen complaints—are likely to be intensified in urban areas. These two environments have different capacities to address the impacts of oil and gas extraction and also experience the impacts and costs in different ways.

VISIBLE AND INVISIBLE PUBLIC COSTS

One of the most visible costs of UGE relates to the transport of inputs—water, chemicals, heavy equipment, and construction materials—to the drilling sites, and outputs, particularly contaminated waste from the drilling site. Well over a thousand truck trips, many by very heavy vehicles, are required to service one well site. These trucks travel over long distances and negatively affect state as well as local road infrastructure.

In the states in which estimates are available, the cost of maintaining local roads exceeds the amount received from oil and gas severance taxes.[43] In Texas, for example, in 2012–2013 it was estimated that the state would receive $3.6 billion in severance taxes from oil and gas production; however, the Texas Department of Transportation estimated that the damage to roads from drilling operations totaled approximately $4 billion.[44]

Costs related to public safety are not inconsequential. An analysis of six states with UGE operations found that traffic accidents quadrupled with a concomitant increase in fatalities.[45] Jacquet, too, found that in Sublette County, Wyoming, traffic on roads increased, as did the number of traffic accidents, the number of emergency room visits, and the demand for emergency response services.[46] Increases in traffic accidents and citations of commercial vehicles have also been documented in Ohio counties where UGE industrial development has occurred. Ten of the 14 counties experienced increases that exceed the state average rates. In combination with statistics on crime in these same counties, the shale gas development counties were found to be less safe than they were before UGE development.[47] Using the Uniform Crime Statistics database, James and Smith, in a multicounty study of shale gas development regions in the United States, found that between 2000 and 2012 UGE counties experienced faster growth in reported crimes, including violent crimes, than the U.S. average.[48]

IMPACTS BEYOND THE DRILLING REGIONS

Although there are demonstrated local or regional impacts of UGE, there are more distant impacts that also need to be addressed. UGE requires quantities of chemicals, sand, and water, which have to be trucked to the drilling site. With the exception of water, these products generally come from areas far from drilling sites. The proliferation of sand mining in Western Wisconsin and Eastern Minnesota, for example, has transformed what were once small rural towns whose economies centered on agriculture and tourism to mining centers providing sand (silica) to distant drilling sites. The local population in these centers is divided over the environmental and public health hazards of silica dust.[49–51]

Fracking fluid, wastewater, and other liquid waste, byproducts of the UGE process, must also be disposed of safely. The current primary method for disposal of contaminated water is the injection well, a bored or drilled shaft that inserts fluid deep underground into porous rock formations. Like many of the environmental, health, and safety issues associated with UGE, the role of injection wells needs to be understood in the broader context of the extraction process as a whole. The United States has approximately 680,000 waste and injection wells for disposal of hazardous waste. Although this method of disposal has been used for decades, the 21 billion barrels of contaminated water produced in a year clearly exceeds the current supply of disposal sites. Furthermore, many of the existing injection wells are located far from the drilling sites. For example, two of Pennsylvania's injection wells are located in the far northwestern corner of the state on the border with New York State while the overwhelming majority of wells in Pennsylvania are located in the central, northeastern, and southwestern part of the state.

Much of Pennsylvania's contaminated water is also trucked to neighboring Ohio's injection wells. In 2012, Ohio injection wells handled 588 million gallons of wastewater, the majority of which was received from Pennsylvania.[52] However, Ohio is beyond capacity to handle wastewater from Pennsylvania. The disposal of toxic waste from UGE in Pennsylvania and Ohio may include a wider range of far-flung sites including injection wells in Gulf Coast states. Waste materials will be transported to these injection wells via barges on the Mississippi River. Because of the search for new locations for unwanted drilling outputs, one of the most widely geographically distributed products of UGE development may be toxic waste.

Although the oil and gas industry has emphasized trends to recycle wastewater, the well servicing businesses have been unable to find a technology that will enable them to profit from recycling and the recycling alternative is largely considered a failure. Less than 10% of shale oil and gas field well water is recycled, and there is no expectation that this will increase.[53]

Wherever they are located, injection wells hold their own distinctive risks and environmental costs. For example, injection wells located in central and eastern U.S. states are now being linked to increased seismic activity in regions of the United States that previously experienced few earthquakes (e.g., Ohio).[54] Recent scientific evidence on humanly induced seismicity indicates that injection wells may be associated with earthquakes many miles (up to 50 kilometers) from the wellbores.[55] The hazards from induced seismicity can have an impact on dams, nuclear power plants, and other critical facilities.

A second geographically dispersed risk associated with byproducts from the drilling process is that of solid waste, deep underground tailings and sludge byproducts. The tailings may have unusually high concentrations of naturally occurring radiation, particularly ^{226}radium, which implies an unusual disposal risk for landfills, the overwhelming majority of which cannot manage hazardous waste safely.[56] In addition to receiving millions of gallons of wastewater, Ohio also receives hazardous waste from UGE for disposal. Perhaps ironically, tailings from Pennsylvania are also being transported to New York State despite the fact that UGE is not permitted in that state. Solid waste is deposited in landfills in multiple New York counties that were once used for local waste deposit but now have become major deposit sites for hazardous waste produced at the Pennsylvania drilling sites.[57]

The evidence concerning present and future contaminated waste disposal demonstrates that risk-bearing activities associated with the drilling process may occur far from the drilling sites, including in areas where drilling is highly regulated or prohibited. Furthermore, as the Ohio case demonstrates, disposal risks are being concentrated in particular regions creating another set of distinctive risks, different from those created in drilling sites. Thus, the geography of UGE is complex, with particular states and regions bearing more differentiated risks than others.

One reason for the distribution of UGE-related activities is geologic. Some areas offer better conditions for containing UGE waste. Another reason explaining the UGE footprint, however, is fragmentation and differentiation of regulations governing the interrelated activities in the UGE process (see Sinding et al., Chapter 9, this volume). This fragmentation allows for "venue shopping" to find less regulated locations for the disposal of toxic wastes.

In summary, this chapter has tried to show that the footprint of UGE is local, regional, and national. The drilling may be in one small rural area, but the ripple effect goes far beyond that. Although the well pad may be the locus of production, the environmental and economic costs of servicing the well site are distributed in a complex production chain that stretches across the United States.

NOTES

1. Wirfs-Brock J. Groundwater, Noise, Odors Top Colorado Oil & Gas Complaints. *Inside Energy*. http://insideenergy.org/2014/07/01/groundwater-noise-odors-top-colorado-oil-gas-complaints (accessed July 10, 2014).

2. Cortese CF, Jones B. The sociological analysis of boomtowns. *Western Sociological Review* 8, no. 1 (1977): 75–90.

3. Freudenburg WR, Wilson LJ. Mining the data: analyzing the economic implications of mining for nonmetropolitan regions. *Sociological Inquiry* 72 (2002): 549–557.

4. Kassover J, McKeown RL. Resource development, rural communities and rapid growth: Managing social change in the modern boomtown. *Minerals and the Environment* 3, no. 1 (1981): 47–57.

5. Freudenburg WR, Gramling R. Economic linkages in an extractive economy. *Society and Natural Resources* 11 (1998): 569–586.

6. Marchand J. *Local Labor Market Impacts of Energy Boom Bust Boom in Western Canada*. Edmonton, Canada: Department of Economics, University of Alberta, 2011. Available from the author.

7. Weinstein A, Partridge M. *The Economic Value of Shale Natural Gas in Ohio*. The Swank Program in Rural-Urban Policy Summary and Report, Department of Agricultural, Environmental, and Development Economics, The Ohio State University, Columbus, 2011. http://aede.osu.edu/research/c-william-swank-program/policy-briefs (accessed December 30, 2013).

8. Weber J. The effects of a natural gas boom on employment and income in Colorado, Texas, and Wyoming. *Energy Economics* 34 (2012): 1580–1588.

9. Bureau of Labor Statistics. *Quarterly Census of Employment and Wages for NAICS 211: Oil and Gas Extraction Industries for 2002 and 2012*. http://data.bls.gov/cgi-bin/dsrv?en (accessed July 30, 2014).

10. Ibid.

11. Haggerty J, Gude P, Delorey M, Rasker R. *Long Term Effects of Income Specialization in Oil and Gas Extraction in the U.S. West, 1980–2011*. Bozeman, MT: Headwaters Economics, 2014. http://headwaterseconomics.org/wphw/wp-content/uploads/OilAndGasSpecialization_Manuscript_2013.pdf (accessed August 22, 2014).

12. Best A. Bad gas or natural gas: the compromises involved in energy extraction. *Planning Magazine* (October 2009): 30–34. http://www.planning.org/planning/2009/oct/ (accessed December 30, 2013).

13. Rigs World 2011. 2014. http://www.rigsworld.com/Rig-Count.htm (accessed August 22, 2014).

14. Feser E, Sweeney S. *Out-Migration, Population Decline, and Regional Economic Distress*. Report prepared for the U.S. Economic Development Administration. Washington, DC: U.S. Department of Commerce. 1999.

15. Christopherson S, Rightor N. How shale gas extraction affects drilling localities: lessons for regional and city policy makers. *Journal of Town and City Management* 2 (2012): 350, 358–366.

16. Feser and Sweeney, *Out-Migration*, op. cit.

17. Weinstein and Partridge, *The Economic Value of Shale Natural Gas in Ohio*, op. cit.

18. Jacquet JB. *Workforce Development Challenges in the Natural Gas Industry* (Working Paper Series for "A Comprehensive Economic Impact Analysis of Natural Gas Extraction in the Marcellus Shale"). Ithaca, NY: Department of City and Regional Planning, Cornell University, 2011. http://www.greenchoices.cornell.edu/development/marcellus/reports.cfm.

19. Ibid.

20. Office of the New York State Comptroller. Growth in local sales tax collections slows to 2.4 percent in first half of 2014. Long Island and Southern Tier Collections Decline. http://www.osc.state.ny.us/localgov/pubs/research/snapshot/localsalestaxcollections0714.pdf. (accessed August 24, 2014).

21. Ecosystem Research Group. *Sublette County Socioeconomic Impact Study, Phase I Final Report*. Prepared for the Sublette County Commissioners. Missoula, MT: Ecosystem Research Group, 2008.

22. Jacquet JB. *Energy Boomtowns and Natural Gas: Implications for Marcellus Shale Local Governments and Rural Communities* (NERCRD Rural Development Paper No. 43). University Park, PA: Northeast Regional Center for Rural Development, The Pennsylvania State University, 2009.

23. Holeywell R. North Dakota's oil boom is a blessing and a curse. *Governing* (August 2011). http://www.governing.com/topics/energy-env/north-dakotas-oil-boom-blessing-curse.html (accessed August 22, 2014).

24. Jacquet, *Energy Boomtowns and Natural Gas*, op. cit.

25. Holeywell, North Dakota's oil boom, op. cit.

26. McPherson J. Oil boom raises rents in ND, pushes seniors out. Associated Press, November 14, 2011. http://www.nbcnews.com/id/45292393

/ns/us_news-life/t/oil-boom-raises-rents-nd-pushes-seniors-out/#.Uffox OBh5UQ (accessed July 26, 2013).

27. Weinstein and Partridge, *The Economic Value of Shale Natural Gas in Ohio,* op. cit.

28. Holeywell, North Dakota's oil boom, op. cit.

29. Christopherson S, Rightor N. *How Should We Think about the Economic Consequences of Shale Gas Drilling?* (Working Paper Series for "A Comprehensive Economic Impact Analysis of Natural Gas Extraction in the Marcellus Shale.") Ithaca, NY: Department of City and Regional Planning, Cornell University, 2011. http://www.greenchoices.cornell.edu/downloads/development/shale/marcellus/Thinking_about_Economic_Consequences.pdf (accessed July 26, 2013).

30. Navarro M. With gas drilling next door, county in New York gets an economic lift. http://www.nytimes.com/2011/12/28/nyregion/hydrofracking-gives-chemung-county-ny-economic-boost.html?pagewanted=all&_r=0 (accessed August 22, 2014).

31. New York State Office of the Comptroller. *Local Government Snapshot.* February 2014. http://www.osc.state.ny.us/localgov/pubs/research/snapshot/localsalestaxcollections0214.pdf. (accessed August 22, 2014).

32. Wilber T. *Under the Surface: Fracking, Fortunes, and the Fate of the Marcellus Shale.* Ithaca, NY: Cornell University Press, 2012.

33. Oldham J. *North Dakota Oil Boom Brings Blight with Growth as Costs Soar.* http://www.bloomberg.com/news/2012–01–25/northdakota-oil-boom-brings-blight-with-growth-as-costs-soar.html (accessed October 20, 2013).

34. McGowan E. Gas drilling turning quiet tourist destination into industrial town (one of a seven-part series by the author). *InsideClimate News,* May 20, 2011. http://insideclimatenews.org/news/20110517/fracking-pennsylvania-natural-gas-drilling-marcellus-shale (accessed October 10, 2013).

35. Rumbach A. *Natural Gas Drilling in the Marcellus Shale: Potential Impacts on the Tourism Economy of the Southern Tier* (Working Paper Series for "A Comprehensive Economic Impact Analysis of Natural Gas Extraction in the Marcellus Shale"). Ithaca, NY: Department of City and Regional Planning, Cornell University, 2011. http://greenchoices.cornell.edu/development/marcellus/policy.cfm (accessed November 13, 2014).

36. Adams R, Kelsey, T. *Pennsylvania Dairy Farms and Marcellus Shale, 2007–2010* (Pennsylvania State University Extension Marcellus Fact Sheet). University Park, Pennsylvania, 2011. http://pubs.cas.psu.edu/FreePubs/PDFs/ee0020.pdf (accessed October 15, 2013).

37. National Resources Conservation Service, U.S. Department of Agriculture. *Reducing the Impact of Natural Gas Compressor Noise.* http://www.pa.nrcs.usda.gov (accessed August 20, 2014).

38. Finger R. Watkins Glen board opposes gas storage plans. *Star Gazette,* August 20, 2014. http://www.stargazette.com/story/news/local/2014/08/20/watkins-glen-village-board-opopsition-gas-storage-plans/14338571 (accessed August 23, 2014).

39. U.S. Bureau of the Census. *Energy Boom Fuels Rapid Population Growth in Parts of Great Plains; Gulf Coast Also Has High Growth Areas, Says Census Bureau.* http://www.census.gov/newsroom/releases/archives/population/cb14-51.html (accessed August 23, 2014).

40. Ibid.

41. Ibid.

42. Ibid.

43. Batheja A. Cash for road repair in shale areas proves elusive. *The Texas Tribune*, April 26, 2013. http://www.texastribune.org/2013/04/26/cash-for-road-repair-in-shale-areas-proves-elusive (accessed August 23, 2104).

44. Ibid.

45. Begos, K. Traffic accidents are a deadly side effect to fracking boom. Associated Press. http://newsok.com/traffic-accidents-are-a-deadly-side-effect-to-fracking-boom/article/4746478. 2014. (accessed 23 August, 2014).

46. Jacquet, *Energy Boomtowns and Natural Gas*, op. cit.

47. Auch T. *Crime and the Utica Shale*. http://www.fractracker.org/2014/06/crime-utica-shale (accessed August 5, 2014).

48. James A, Smith B. *There Will Be Blood: Crime Rates in Shale-Rich US counties*. Oxford Centre for the Study of Resource Rich Economies, Oxford University, 2014. http://alexandergjames.weebly.com/uploads/1/4/2/1/14215137/crime.resources.pdf (accessed July 31, 2014).

49. Karnowski S. Natural gas, oil boom spurs sand mining in Midwest. *USA Today*, January 8, 2012. http://usatoday30.usatoday.com/money/industries/energy/story/2012-01-08/fracking-boom-sand-mining/52398528/1 (accessed August 23, 2014).

50. Deller SC, Schreiber A. *Frac Sand Mining and Community Economic Development* (Staff Paper Series, Number 565). Madison, WI: Department of Agricultural and Applied Economics, University of Wisconsin, May 2012.

51. Department of Natural Resources, State of Wisconsin. *Analysis of the Petition for Promulgation of Rules for Respirable Crystalline Silica*. http://fracsandfrisbee.com/wp-content/uploads/2012/04/Final-Silica-Petition-Response-01-30-12.pdf (accessed August 24, 2014).

52. Lutz B, Lewis A, Doyle M. Generation, transport, and disposal of wastewater associated with Marcellus Shale gas development. *Water Resources Research* 49 (2013): 1–10.

53. Ibid.

54. Choi C. Fracking practices to blame for Ohio earthquakes. *Live Science*, September 4, 2013. http://www.livescience.com/39406-fracking-wasterwater-injection-caused-ohio-earthquakes.html (accessed August 22, 2014).

55. Keranen K. Potentially induced earthquakes in Oklahoma, USA: Links between wastewater injection and the 2011 Mw 5.7 earthquake sequence. *Geology* (July 2014): G34045.1.

56. Belcher M, Resnikov M. Hydraulic fracturing radiological concerns for Ohio. 2013. http://www.slideshare.net/MarcellusDN/hydraulic-fracturing-radiological-concerns-for-ohio?related=1 (accessed August 23, 2014).

57. Mantius P. *New York Imports Pennsylvania's Radioactive Waste Despite Falsified Water Tests*. National Security News Service. August 13, 2013. http://www.dcbureau.org/201308148881/natural-resources-news-service/new-york-imports-pennsylvanias-radioactive-fracking-waste-despite-falsified-water-tests.html (accessed August 22, 2014).

Chapter 9

The Regulation of Shale Gas Development

Kate Sinding, Daniel Raichel, and Jonathon Krois

INTRODUCTION

Consistent with the American system of federalism, the regulation of shale gas development in the United States is divided among multiple layers of government: federal, state and local, and, to varying extents, from state to state. Unlike most industrial activities, however, shale gas extraction is regulated first and foremost by the states, rather than the federal government. Because of a network of exemptions from many of the nation's major federal environmental laws, federal agencies exercise authority over discrete elements of the process with hydraulic fracturing expressly excluded from regulation.

With shale gas extraction currently underway in states ranging from the Northeast through the Midwest, the South and the western United States, the result is a patchwork of state laws that vary widely in their scope, stringency, and enforcement. States govern, to differing degrees, virtually every aspect of shale gas development, from exploration through design, location, construction, operation and, ultimately, abandonment. In addition, states regulate to address the wide array of potential environmental impacts associated with shale development, including those relating to water, air, wildlife, habitat, and waste.

The extent to which local governments (e.g., counties, cities, and towns) can exercise control over gas development activities within their borders varies. Local control, itself a creature of state law, ranges from broad authority to regulate or even ban shale gas development in some states to a virtual absence of authority to affect the activity through local laws in other states. This chapter presents an overview of the legal and regulatory aspects of shale gas development.

REGULATION OF SHALE GAS DEVELOPMENT AT THE FEDERAL LEVEL

The role of the federal government in regulating shale gas development is actually relatively limited, as the states act as the primary regulators. This is due to a network of interrelated exemptions from the nation's bedrock environmental laws, including the exemption of hydraulic fracturing from regulation by the leading federal environmental regulator, the Environmental Protection Agency (EPA). These exemptions are not absolute, however, because the federal government does retain the authority to regulate a number of important aspects of overall shale gas development. In some instances, implementation and enforcement of these federal laws is delegated to the states. The following presents a brief overview of some of the major laws that impact the regulation of shale gas development.

Federal Exemptions for Shale Gas Development

Safe Drinking Water Act

Probably the most famous federal exemption is that which precludes the EPA from regulating hydraulic fracturing under the Safe Drinking Water Act (SDWA).[1] The SDWA was enacted in 1974 to protect public drinking water supplies as well as their sources. Under the act, the EPA is authorized to establish health-based standards for drinking water to protect against both naturally occurring and manmade contaminants.

The SDWA's Underground Injection Control (UIC) program regulates how industrial and municipal waste and other fluids can be injected into underground strata that contain, or may be a source of, drinking water.[2] The act regulates the entire injection operation, including permitting, siting, construction, operation, maintenance, monitoring, testing, and closing of underground injection sites. The EPA and the states jointly implement the act to ensure protection of these standards, and states are able to obtain so-called primacy, that is, primary regulatory authority, from the EPA over oil- and gas-related UIC wells.

In 2005, following a highly controversial 2004 EPA report[3] that found that chemicals used in hydraulic fracturing "pose little or no threat" to drinking water, Congress expressly exempted hydraulic fracturing (except that using diesel fuels) from regulation under the UIC program.[4] The injection of natural gas for storage is also expressly exempted from regulation under the SDWA.[5] As a consequence, the EPA has no authority to promulgate rules that govern this fundamental aspect of shale gas extraction. That said, and as addressed later in the chapter, the UIC program does govern the subsurface injection of oil and gas production wastewaters.

Clean Water Act

Preventing pollution of surface water bodies is the primary objective of the Clean Water Act (CWA).[6] Although parts of the CWA do apply to shale gas development, oil and gas exploration and production operations are generally exempt from the CWA's stormwater discharge permit program. As long as runoff at drilling sites, well pads, and transmission corridors are deemed uncontaminated, the industry does not need to comply with the program's permitting requirements, irrespective of the size of the operation. In 2005, the category of exempted activities was expanded to include all field activities and operations, including new roads and pipelines associated with oil and gas production.[7]

In addition, hydraulic fracturing fluids used in shale gas production are exempted from the definition of pollutants that are subject to permitting under the CWA's National Pollutant Discharge Elimination System (NPDES) Program.[8] The NPDES Program makes it unlawful to discharge any pollutant from a so-called point source into the navigable waters of the United States without an approved permit. This exemption also extends to produced water that is "disposed of" by reinjection into gas production wells.

Clean Air Act

First enacted in 1970, then significantly amended in 1977 and 1990, the Clean Air Act (CAA) limits emissions of both hazardous pollutants and other common air pollutants that can present a risk to human health, for example, carbon monoxide and lead (known as "criteria pollutants").[9] As discussed in other chapters, oil and gas wells are significant sources of volatile organic compounds (VOCs), hydrogen sulfide, and other hazardous air pollutants, and criteria pollutants are emitted from trucks and other machinery used in oil and gas production. The National Emission Standards for Hazardous Air Pollutants (NESHAPs) is, as suggested by its name, the section of the CAA that regulates 190 identified pollutants known or suspected to cause cancer or other serious health effects.[10] Exemptions from the CAA's hazardous air pollutant requirements, however, mean that wells and well fields cannot be regulated either collectively or individually as major sources, or individually as small so-called area sources.[11]

In another concession to the oil and gas industry, hydrogen sulfide, which is commonly associated with oil and gas production, was not listed by Congress as a hazardous air pollutant in the CAA even though human exposure is linked to irritation, difficulty breathing, nausea, vomiting, headaches, loss of consciousness, and even death.[12] Other exemptions exist under the CAA's Prevention of Significant Deterioration Program, which limits incremental increases of air pollutants in areas where air quality exceeds national standards, and the Nonattainment Area Program, which imposes additional permitting requirements on sources in areas in which national standards are regularly exceeded.[13,14] As discussed subsequently, other aspects of oil and gas production are subject to parts of the CAA.

Resource Conservation and Recovery Act and Superfund

Shale gas development produces several types of waste including drill cuttings, drilling fluids, produced water, and flowback water. Much of this waste is dangerous to human health and the environment, containing harmful hydraulic fracturing fluid chemicals as well as toxic naturally occurring substances.[15] The Resource Conservation and Recovery Act (RCRA) is the primary federal law designed to ensure the safe management of solid and hazardous wastes "from cradle-to-grave"—through generation, transportation, treatment, storage, and disposal—and thus to prevent the creation of new toxic waste sites.[16] Notwithstanding that wastewater and solid wastes generated through the exploration and production of oil and gas can contain substantial quantities of contaminants, these wastes are categorically exempted from the definition of hazardous waste under RCRA.[17] As such, they escape the act's comprehensive scheme for the testing and safe handling and transportation and disposal of hazardous wastes, although they are subject to the less stringent provisions governing nonhazardous wastes.

In addition, RCRA provides that the EPA and citizen plaintiffs can bring cleanup actions against a responsible party whose waste handling practices have created an "imminent and substantial endangerment to health or the environment."[18] Wastes otherwise exempt from the definition of hazardous waste under RCRA may be subject to liability under this exemption (though the responsible party may choose a cleanup option that is less stringent than might be required for hazardous wastes).[19]

Whereas the RCRA governs the management of wastes to avoid risks to human health and the environment, the Comprehensive Environmental Response, Compensation and Liability Act (CERCLA, commonly referred to as "Superfund") provides a framework for the cleanup of toxic waste sites.[20] CERCLA expressly excludes petroleum and natural gas from its definition of "hazardous substance," as well as many other toxic substances such as VOCs, polyaromatic hydrocarbons, arsenic, and mercury, when they occur naturally in oil or gas.[21] This exemption was won by the oil and gas industry in exchange for its commitment to pay into the original cleanup fund established under the act (the Superfund), an obligation that sunset in 1995. Moreover, in adopting the definition of "hazardous waste" under other statutes, including RCRA, CERLCA also inherits RCRA's exemption of oil and gas exploration and production wastes.

Emergency Planning and Community Right-to-Know Act

The Emergency Planning and Community Right-to-Know Act (EPCRA) was enacted in 1986 to establish a process for informing people of chemical hazards in their communities.[22] Covered industries are required to report the locations and quantities of certain chemicals stored, released, or transferred, and some of this information is made available to the public through annual publication of a Toxics Release Inventory (TRI).[23] In the EPA's implementation of EPCRA, oil and gas exploration and production facilities are not required to report to the TRI. However, other provisions of EPCRA governing emergency release notification and reporting do apply to oil and gas well sites.[24,25]

National Environmental Policy Act

Shale gas development on federal lands is regulated by the Bureau of Land Management (BLM) within the Department of the Interior. In general, BLM's regulation of such activities is subject to the comprehensive review provisions of the National Environmental Policy Act (NEPA), which requires the preparation of an environmental impact statement for any major federal approval significantly affecting the environment.[26] As part of the 2005 Energy Policy Act, a so-called categorical exclusion from NEPA review was enacted that applies to certain, not uncommon, oil and gas exploration and development activities.[27]

Existing Sources of Federal Regulatory Authority over Shale Gas Development

Although the network of exemptions just described is substantial, the federal government does regulate shale gas development in some important ways. Moreover, there are existing sources of federal authority that are not currently being fully exercised but could be used to increase federal control over the environmental impacts of oil and gas production.

Water Quality

Although the discharge of hydraulic fracturing fluids is exempt from regulation under the CWA NPDES program, the EPA does regulate other point source discharges associated with oil and gas development under that program. Facilities that handle the treatment and disposal of produced water, for example, must obtain NPDES permits from the EPA or from the state if the program has been delegated. In 2011, the EPA announced that it was developing new discharge standards (so-called effluent limitation guidelines) for facilities handling wastewater from oil and gas production,[28] although they remain to be issued for public comment.

Hydraulic fracturing is generally exempt from federal regulation under the SDWA, except where diesel fuel is used in fluids or propping agents. The EPA has not promulgated regulations governing this activity but has issued a guidance document. Specifically, injection of wastewaters from oil and gas production is regulated under the SDWA's UIC Program. However, because of the RCRA exemption from the definition of hazardous waste for oil and gas production wastes, the wells into which such wastewater is injected are regulated as so-called Class II wells, which are subject to less stringent regulatory requirements than those that apply to the injection of hazardous wastes into Class I wells. As under RCRA, the SDWA's "imminent and substantial endangerment" provision may be used to address threats related to hydraulic fracturing and other oil and gas development activities.[29]

Limiting emissions of air pollutants applies to a variety of shale gas development facilities. Despite the major exemption for aggregating sources of hazardous air pollutants discussed earlier, NESHAPs do exist for the oil and gas industry, albeit limited to those individual facilities that have the potential to emit 10 tons or more per year of a hazardous air pollutant or 25 tons or more per year of a combination of pollutants, as well as for area sources, which are sources of hazardous air pollutants that are not defined as major sources.[30]

A critical provision under the CAA is the New Source Performance Standards (NSPS), which apply to new stationary facilities or modifications to stationary facilities that result in increases in air emissions.[31] In 2012, the EPA promulgated NSPS for the oil and gas industry that require reductions of VOC emissions at oil and gas well sites, including wells using hydraulic fracturing.[32] In April 2014, the EPA issued a series of white papers regarding methane and VOC emissions in the oil and gas sector that might presage the adoption of new NSPS specifically geared to methane as well as new rules that would limit emissions of pollutants from existing (as opposed to new or modified) facilities.[33]

Federal Lands

As stated previously, the federal government, through the BLM, plays the dominant role in regulating oil and gas development on federal lands. BLM is tasked with leasing subsurface mineral rights not just for the land BLM controls directly but also for lands controlled by other federal agencies, including those managed by the U.S. Forest Service.[34] Its authority to regulate mineral extraction extends to minerals beneath BLM lands, national parks and forests, national wildlife refuges, Indian lands, as well as to minerals in federal ownership beneath privately owned lands.[35] The BLM has promulgated rules that govern all aspects of oil and gas development, and it recently issued proposed revisions to update its rules and address specific issues related to the expansion of hydraulic fracturing.[36]

Other Federal Laws

The Toxic Substances Control Act (TSCA) authorizes the EPA to regulate the manufacture, processing, use, distribution in commerce, and disposal of chemical substances and mixtures.[37] Under TSCA, the EPA maintains a list of chemicals that are or have been manufactured or processed in the United States called the TSCA inventory (which currently contains more than 84,000 chemicals).[38] Because of issues associated with the lack of disclosure of chemical use in hydraulic fracturing, it is unknown how many chemicals used in shale gas development are listed on the inventory. The EPA is currently evaluating that question on the basis of information provided by some companies as part of its ongoing study regarding the risks of hydraulic fracturing to drinking water supplies. It has not, however, used its authority under TSCA to more comprehensively regulate chemical disclosure at the federal level.

Other federal environmental and public health laws also apply to the oil and gas industry, including the Endangered Species Act, which protects animals and plants listed as "endangered" or "threatened,"[39] and the Occupational Safety and Health Act, under which specific standards have been established to promote the health and safety of workers in the oil and gas industry.[40]

Regulation of Shale Gas Development at the State Level

Although the federal government has some regulatory authority, the states have the primary responsibility for regulating oil and gas development in the United States. Most significant regulation of shale gas development comes from state law, and the states have nearly unfettered authority to regulate it. Given the complex, controversial, and risky nature of oil and gas extraction, as well as its putative economic benefits, states have developed rules and regulations to oversee many aspects of the industry from beginning to end. States have also developed protective regulations aimed at controlling for the potentially significant negative externalities of oil and gas development. This section first looks at those parts of the development process states can and do regulate. Second, it addresses those protections against the negative aspects of shale gas development that some states have adopted. Finally, it discusses regulation of other legal issues surrounding gas development.

Testing for Gas

Drillers looking to develop a shale gas well must first locate productive areas of gas. Two commonly used techniques to locate productive areas are seismic testing, which involves using explosives or heavy equipment to create vibrations that enable drillers to identify different types of underground materials and drilling test wells at a particular location.[41] Perhaps not surprisingly, these processes pose some environmental and safety concerns. Seismic testing risks surface damage and damage to underground aquifers. Unplugged test wells and holes for explosives pose basic safety risks and could potentially allow for pollutants to get underground. In response to these risks, there are several controls states impose on operators during this phase; for example, permits and/or blaster's licenses are commonly required before using explosives or conducting other seismic testing. Additional rules vary widely depending on the state. Those protections include requiring operators provide notice to relevant state agencies, oversight by those agencies, minimum setbacks from sensitive areas, and posttesting plugging requirements.[42]

Location

The location of access roads, well pads, and waste disposal sites is often a subject of state regulation. This regulation is usually expressed as forbidding operations within a set distance from other land uses, other oil and gas development, or certain sensitive areas. Much regulation in this area focuses on well spacing—for example, where wells are located in relation to other wells. These requirements often predate the modern expansion of hydraulic fracturing. The motivation for this kind of regulation is primarily economic: the regulations are designed to avoid waste and to foster the efficient development of the natural resource being extracted.[43]

Another form of regulation, setback requirements, is designed to protect other nearby property owners and sensitive environmental resources. Concerned about the risks of blowouts and spills, most shale gas states require that well pads, surface pits, and/or certain methods of drilling and disposal be located a minimum distance away from designated areas. These areas often include residences, schools, property lines, bodies of water, sensitive habitats, water wells, and public water supplies. The average setback is 308 feet, although regulations vary greatly and can range from 100 feet to 1,000 feet.[44]

There is broad agreement among the states that setbacks are appropriate. That said, there is much disagreement about how far is "far enough" and about whether any setbacks are sufficient to protect human health. Making this debate more difficult is a lack of reliable scientific data on the effectiveness of setbacks.[45] A Duke University study recently found evidence of water contamination up to 1 kilometer from drilling sites—far beyond the setbacks usually required.[46]

Casing and Cementing

Casing and cementing the well bore are the primary methods operators use to maintain well integrity and are also common to traditional oil and gas extraction. States regulate casing and cementing to protect underground water supplies. To avoid contamination, well casing must prevent the migration of substances out of the well during drilling and hold under pressure during fracturing. Well casings also help prevent the well from leaking in the future, after it is no longer producing. However, many states wrote their regulations on casing and cementing before the recent explosion in high-volume hydraulic fracturing. Although most producing states have requirements for the minimum depth required when casing a well, few meet the industry best practice of casing at least 100 feet below the deepest underground source of drinking water.[47] Other states express these requirements as generalized demands that casing be sufficient to protect groundwater.[48]

States regulate many other aspects of casing and cementing. For example, requiring surface casing is nearly ubiquitous. Several states also regulate the type of cement used, whether by requiring a specific composition or specific qualities. Pressure testing casing before installation is a requirement in a handful of states. North Dakota, for example, requires operators of a new well to demonstrate its "mechanical integrity" through testing.[49] Some states, but not all, require operators to maintain comprehensive casing logs. Others require that a regulatory agency staffer familiar with casing be present when well casing occurs.

The expansion of casing and cementing regulations is a topic of major concern. Many of the objections to hydraulic fracturing from landowners, environmentalists, and public health advocates relate to concerns about groundwater contamination. Casing and cementing is designed to prevent that contamination. When industry acknowledges the risk of contamination, they often point to defective casing and cementing as the culprit. Much debate, as a result, has focused on how to oversee this process.

Fracturing the Well

An increasing number of states are beginning to regulate the hydraulic fracturing process in greater detail, primarily focusing on the prevention of spills and blowouts and on ensuring the state's ability to respond. States also regulate the fracturing process by regulating the transport and on-site storage of hydraulic fracturing and drilling fluids and of fracking waste. A limited number of states also require operators to notify the relevant state agencies before hydraulic fracturing and after its completion.[50] Many of these regulations, including waste storage, disposal, addressing the use of water, and chemical disclosure requirements, are discussed in the following subsections.

On-Site Waste Storage

Waste from the drilling process is typically stored on site in pits or tanks before being disposed of permanently. States regulate storage to prevent these harmful wastes from contaminating the soil, nearby bodies of water, groundwater, and animals. Producing states almost always regulate the storage in open pits, with some states requiring sealed storage for at least some types of fluid. No state mandates sealed tank storage for every type of fluid. Specific regulations vary depending on the type of waste being stored.[51] Most states require pit liners for at least some pits, as well as requiring freeboard—a distance between the top of the pit and its maximum fluid level.

Wastewater Disposal

Disposal of wastewater from shale gas development is difficult and dangerous. Research has shown that currently available disposal methods may be inadequate and that improper handling, treatment, and disposal of wastewater can expose people and wildlife to toxic, radioactive, or carcinogenic chemicals.[52] This difficulty is compounded by volume; initially, a well can return as much as 100,000 gallons of flowback water per day for several days. These fluids are generally stored on-site in storage tanks and waste impoundment pits before treatment or disposal.

One method of disposal is to discharge water into surface waters after treatment at a wastewater treatment facility. However, flowback water can pose challenges for treatment facilities that are generally unable to remove radioactive and other harmful materials, as well as large amounts of sodium, chloride, and bromide. Another method of disposal is returning flowback water underground using an underground injection well, which is permitted in 30 states. This method of disposal is controversial, however, because it has been linked to earthquakes.[53]

Site Restoration

Most states require the operator to comprehensively "restore" the site of a shale gas operation after the driller is finished. Usually, operators are required to remove contaminated soils, empty waste pits and fill them in, and stabilize the soils on site.

Controlling the Risks of Shale Gas Development

Testing and Replacing Water Supplies

Testing water supplies before and after drilling is highly valuable, and can be critical, in determining whether shale gas development activity has caused contamination. Many state oil and gas laws require operators to restore or replace a contaminated water supply or to compensate the affected landowners if the contamination is the result of the operator's activity. The majority of states, however, do not require operators to test nearby water wells *before* drilling. Pennsylvania, somewhat uniquely, provides for a rebuttable presumption that contamination found within 2,500 feet of wells and within 1 year of drilling is attributable to the operator, unless the operator can present evidence of preexisting contamination.[54]

Addressing Air Emissions during Drilling and Fracturing

Well site equipment emits smog-forming VOCs; known carcinogens, such as benzene; methane; and other air pollutants into the atmosphere. Air pollution is primarily regulated at the federal level under the CAA, but because the CAA primarily focuses on "major" sources much of this federal regulation does not apply to unconventional gas development. Oil and gas operations are generally minor sources that are regulated (if regulated at all) by state "minor" source programs. That said, hydraulic fracturing of shale can cause a number of negative impacts to air quality, yet air emissions are generally underregulated.

The CAA permits states to have air pollution regulations that are more stringent than those in the act itself, although few states actually impose substantive air controls on oil and gas. Some states have used this authority to regulate emissions from drilling activity. These regulations include requirements on capturing VOCs from tanks with the potential to produce them, limitations on simultaneously operating drilling engines, requiring certain kinds of diesel be used in those engines, limitations on the venting of gas, and greenhouse gas mitigation measures. Additionally, a small number of states opt to monitor emissions rather than regulate them directly. Only in one state, Colorado, have regulators approved comprehensive controls on methane emissions from oil and gas development—a response to worsening smog. Colorado's attempt to regulate emissions, however, remains the exception rather than the rule.[55] Most states have left air emissions from shale gas development largely unregulated.

Addressing the Use of Water in Hydraulic Fracturing

Shale gas development using hydraulic fracturing is distinct from other oil and gas development in that it requires the use of massive amounts of water. Water must be withdrawn in large quantities, transported to the site, stored, mixed with chemicals and injected underground, and ultimately recovered as flowback water. Fracturing a single Marcellus Shale well can require between 2.4 and 7.8 million gallons of water. Although states regulate water withdrawals generally, overall regulations have not been written specifically about fracking. Most states require general permits for surface or groundwater withdrawals, and a small number of states regulate water withdrawals by limiting the sources from which water may be withdrawn and by requiring operators to prevent harmful impacts to aquatic ecosystems.

Preventing and Reporting Spills

Spills of hazardous material are a constant risk during the lifetime of a fracked well. Spills can occur when transporting chemicals to well pads; storage pits and tanks may leak or overflow; flowback water and other waste may spill when transported to storage pits; and chemicals may be released during the injection of fracturing fluid into the well, such as during a blowout. State regulations focus on preventing spills, containing them, and requiring reporting and remediation. Most states, at minimum, require operators to have a plan to prevent and control surface chemical spills. Most states also require reporting spills within a certain time frame, although that time frame and the triggering quantities vary. Fewer states require specific spill prevention measures such as dikes, lined pits, and secondary containment features. Some states have gone farther, expanding on CERCLA and regulating the cleanup of certain hazardous chemicals, including some used in hydraulic fracturing.

Other Legal Issues Surrounding State Regulation of Shale Gas Development

Forced Pooling and Property Rights Surrounding Shale Gas Development

States that have traditionally been major centers of oil and gas development often have laws providing for so-called forced pooling, sometimes called "compulsory integration." These laws permit operators who control a certain percentage of the oil and gas rights within a proposed "pool" or "unit" of land to compel the owners of the remaining oil and gas in that unit to lease it the operator for extraction.[56] Forced pooling laws developed out of concerns about waste: traditional oil and gas reserves can become difficult or impossible to extract if too many wells attempt to exploit a single formation. These laws are increasingly being used to facilitate hydraulic fracturing in areas with reluctant landowners. By nature, forced pooling laws permit operators to compel property owners to contract with operators against the owners' will and as such are controversial. Landowners, however, have so far had little success in opposing forced pooling.

State Mini-NEPAs

States also subject many aspects of shale gas development to state environmental review statutes often referred to as "mini-NEPAs." As discussed earlier, NEPA is the federal statute that requires federal agencies to evaluate many of their proposed actions to determine whether they pose a risk to the environment. If that action poses a significant risk of adversely affecting the environment or human health, the agency must thoroughly review the potential impacts, the steps taken to mitigate those impacts, and alternatives to the action in an environmental impact statement. Many states have drawn inspiration from NEPA and impose similar requirements on state agencies, often including municipalities. These mini-NEPAs can require environmental review of oil and gas development-related decisions ranging from permitting to zoning.

Disclosure Requirements

Disclosure rules vary greatly from state to state, and only about half of producing states have fracking disclosure requirements. Among those that do, requiring operators to identify the chemical additives in fracking fluid is relatively common, although far fewer of those states also require operators to reveal the chemicals' concentrations. Chemical disclosure is regulated by the BLM for fracking on federal lands, but the EPA has not used its authority to more broadly regulate disclosure. Some states do regulate disclosure, which can serve a variety of important purposes, such as giving nearby communities advance notice of potential impacts—which allows for more targeted baseline testing—or providing information necessary to evaluate the impacts of fracking, such as locations of well sites, volumes of fluid used, and that fluid's composition.[57] Also relatively common is requiring operators to identify nearby water wells before fracking. Prior notice to nearby residents that a fracking operation will occur, however, is required in only a handful of states. This hodgepodge of regulations generally suffers from poor compliance and enforcement, as well as broad trade secret exemptions.

Regulation of Shale Gas Development at the Local Level

Although the process of shale gas development is often regulated at the state and federal levels, most development (not on federal lands) occurs on land within the jurisdiction of a local government. Not surprisingly, the recent nationwide surge in shale gas production has produced a parallel flood of municipal laws attempting to address local issues related to this new influx, raising the question as to what role local governments may play. Because municipalities are creatures of state constitutional and statutory law, there are at least 50 answers to this question, and in many states it has formed the basis of some of the most pitched legal battles regarding hydraulic fracturing to date. Although the conflict between local and state regulation of shale gas development often revolves around questions of abstract legal interpretation, the factual context looms large. Often at stake is whether local governments are allowed any input over where and how such development occurs in their communities or whether those decisions rest entirely with state officials.

Local Zoning or Land Use Laws

To understand the tension between local and state regulation of shale gas development, it is helpful first to have an understanding of the different ways a local law may affect such activities. By far the most common way is through local zoning or land use laws. In a general sense, zoning laws function by dividing local land into different "zones" based on the permissible use of land (e.g., residential, commercial, industrial) and the intensity of that use (e.g., "light" industrial or "heavy" industrial). A suburban neighborhood zoned "light residential," for instance, may permit single-family homes but exclude factories, commercial businesses, or even denser types of residential development, such as apartment buildings.[58]

Because zoning is almost always comprehensive (i.e., all land within the municipality is placed in a zone), proposed gas development activities may be affected depending on which zone they fall into. For example, assuming local zoning is enforceable, oil and gas operations—likely to be considered industrial activities—would be prohibited from zones restricted solely to residential uses of land. Similarly, a light industrial district may allow for gas development activities, but only where an operator obtains a "conditional use" permit. As the name suggests, conditional use permits (also known as "special use" permits) are a zoning tool that allows local officials to permit certain land uses—usually ones that carry a greater than normal risk of affecting neighborhood character or impairing local development goals—provided particular conditions imposed by the permit are followed.

Zoning-based restrictions may also apply in the absence of a required zoning permit because zones often contain their own general conditions or standards for the use of land. An industrial zone, for instance, may require a minimum distance (i.e., a setback) between new factories and existing buildings or homes. Likewise, a commercial zone may limit the hours of operation for certain businesses like bars or restaurants, or a residential zone may impose noise or light restrictions on residences. In the context of shale gas development, one could imagine an agricultural zone with fencing requirements applicable to well pads or waste pits or a residential zone where operators would need to erect noise dampening walls during drilling and fracturing to meet generally applicable noise standards.

Outside the world of zoning and land use, many municipalities also have laws that apply to broad classes of activities that may also affect shale gas development indirectly—for example, air quality ordinances that require minimum air emissions standards for local industrial facilities or rules respecting the use of heavy trucks on municipal roads. Although the practical distinction between these types of ordinances and zoning laws is not always clear (e.g., there may be little difference between a zoning law containing noise standards and a general "noise ordinance"), the ultimate determination of whether state law overrides or "preempts" a local law may depend on the statutory or constitutional basis of authority that justifies the local law's existence. For example, the Pennsylvania Oil and Gas Act's preemption provision distinguishes between local ordinances passed under the authority of two state laws—the Flood Plain Management Act and the Municipalities Planning Code—and all other local ordinances.[59]

Although more rare, local laws may also try to regulate operational aspects of the development process, directly providing a second layer of technical oversight in addition to that of the state. In Flower Mound, Texas, for example, in order to acquire and maintain a necessary local oil and gas well permit, operators must submit various operational plans for the drill site, such as a hazardous materials management plan; abide by certain operational standards during drilling and fracking; agree to regularly report to and be inspected by the local oil and gas inspector; and post a bond to cover potential catastrophic damage to the town.[60]

Municipal laws may also affect shale gas operations by preventing them entirely. Recently, municipal bans on hydraulic fracturing activities have grown in popularity, with more than 100 such bans (some of which are temporary, known as "moratoriums") having been passed in New York and others now also appearing in Texas, California, Colorado, New Mexico, and Pennsylvania. Bans can take the form of a zoning law (e.g., a town where all land is zoned residential) or can fall under a municipality's more general authority to address potentially hazardous activities. Either way, hydraulic fracturing bans have become a topic of controversy in many states, and no other type of municipal law has as great an impact on shale gas development.

THE LIMIT OF MUNICIPAL REGULATORY AUTHORITY: CONFLICTS WITH STATE STATUTORY LAW

Municipalities are creatures of state law, deriving their power either from state statutory laws or the state constitution directly. Regardless of its source, however, local authority is usually malleable. Where municipalities derive their power from state statutory laws, that power can be limited by other state laws because the state legislature is generally free to amend or repeal its own laws. Similarly, where local power is constitutionally codified, most state constitutions prevent municipalities from enacting laws that conflict with state legislation. There is much debate (and many lawsuits) surrounding the question of whether municipalities can pass laws like the ones discussed in the foregoing sections or, alternatively, whether such laws are allowed to have any effect on shale gas development. The debate mainly centers on the extent to which state oil and gas laws override or preempt local laws. For example, the Ohio constitution grants municipalities general authority over local issues, provided that local laws "are not in conflict with general laws" of the state.[61] Likewise, the Montana Constitution provides that a local government with self-government powers may provide or perform any functions not expressly prohibited by the Montana constitution, law, or charter.[62] Therefore, where a local law does conflict with a state statute, it is generally preempted.

Conflict can be explicit or implicit: a local ordinance may attempt to regulate a subject matter *explicitly* reserved to the state by statute, or it may *implicitly* clash with the purpose or purposes of that statute. In the oil and gas context, explicit preemption is possible when a state oil and gas law provides a "preemption provision" addressing local control. For example, the New York Oil, Gas, and Solution Mining Law states that it "shall supersede all local laws or ordinances relating to the regulation of the oil, gas and solution mining industries."[63] Likewise, explicit limitations on municipal power can come from the statutory source of that power. The Michigan Zoning Enabling Act (the Michigan law that grants that state's municipalities the power to zone), for instance, prevents counties or townships from passing laws regulating the "drilling, completion, or operation of oil or gas wells."[64]

In states where these explicit statutory provisions exist, it is generally clear that municipalities cannot directly target technical operational aspects of oil and gas production. Less clear, however, is whether these provisions actually conflict with local zoning and land use laws, particularly where such laws only indirectly or inadvertently affect shale development activities. A town zoning code may never mention oil and gas activities by name, but town residential zones may nonetheless exclude such activities by default. The question arises, then, whether the explicit preemption provision demands that development be allowed in all zones or whether a distinction can be drawn between a municipality's regulation of land use and the state's regulation of oil and gas operations.

Even trickier questions arise in the field of implied preemption—where courts must determine whether a local law stands as an obstacle to the express or implied purposes of a statute. The North Dakota oil and gas law, for example, which does not have an explicit preemption provision, prevents the state from issuing a permit for a "well that will be located within five hundred feet of an occupied dwelling."[65] Would this provision then preempt a village law in North Dakota requiring a 600-foot setback from an occupied dwelling? Or is the implied purpose of state law only to set *minimum* setback distances, which municipalities can increase to provide additional protection? What if the local setback wasn't from "occupied dwellings" but from another local feature not addressed in the state law, such as property lines? Does a different *kind* of setback conflict with state setbacks for occupied dwellings?

Needless to say, these questions are difficult, and the answers depend on the particularities of both the local and state laws in question. It is therefore not surprising that courts addressing this issue—both in the context of hydraulic fracturing and oil and gas development generally—have come to different conclusions. The Supreme Courts of Colorado and Pennsylvania, for example, have held that their state oil and gas laws did not prevent a municipality from enforcing zoning ordinances specifying where oil and gas activities may or may not take place.[66] Both courts found that traditional local land use powers (e.g., the designation of permissible zones for industrial uses) could be harmonized with the state's exclusive power to regulate oil and gas drilling. In contrast, a lower court in West Virginia came to the opposite conclusion in striking down a local law, holding that the West Virginia oil and gas law comprised a "comprehensive regulatory scheme" with "no exception" for municipal regulatory control over oil and gas wells.[67]

Even in states where it is established that municipalities have some authority over oil and gas activities, the limits of that authority may still vary. Courts in Colorado and New York, for example, have recognized local authority over the location of shale gas development, but to different extents. In New York, the state's highest court, the Court of Appeals, recently recognized the authority of municipalities to completely exclude oil and gas drilling using zoning, declaring that current state law does not "oblige" municipalities "to permit the exploitation of any and all natural resources."[68] In Colorado, by contrast, one such ban was struck down as in conflict with that state's interest in "efficient oil and gas development and production" as expressed in Colorado oil and gas law.[69] Questions can also arise even where local laws broadly allow oil and gas drilling. For example, does the power to zone out hydraulic fracturing imply the power to regulate it where it does occur? In Texas, the nation's top gas-producing state, municipalities have broad discretion both to ban and regulate oil and gas drilling.[70] However, in other states where location-based zoning controls on oil and gas drilling activities have been upheld, open questions remain as to whether municipalities may enforce other traditional zoning controls (such as fencing requirements, noise restrictions, and controls on hours of operation) against shale gas development activities within local borders.

CONSTITUTIONAL LIMITATIONS: A SOURCE OF RESTRAINT AND EMPOWERMENT FOR MUNICIPALITIES

Like ordinary laws passed by the state legislature, the federal and state constitutions can have an impact on whether and how municipalities may pass laws that affect shale gas development. In Pennsylvania, for example, where a municipality completely prohibits an otherwise lawful land use, it may be required to present evidence that the ban legitimately serves a public purpose.[71] Importantly, however, constitutions also constrain the actions of state legislatures and agencies, often putting municipalities and state actors on the same footing. For example, when passing laws or regulations affecting oil and gas development, both local and state decision makers must consider the potential financial impact of constitutional "takings" law. Takings law stems from the Fifth Amendment to the U.S. Constitution, which states that the government cannot "take" private property for a public use without paying "just compensation." Many state constitutions also have "takings clauses," which often provide similar or the same rights as the federal takings clause.[72]

A clear-cut taking occurs when a state or local government uses the power of eminent domain, but the Supreme Court has also recognized that a taking of property may occur, in certain cases, where a law or regulation goes "too far" in restricting the use of private property. In the context of shale gas, there is a question as to whether laws regulating or limiting its development may trigger this so-called takings liability, and litigation has already been brought or threatened in several states against both local and state governments.[73] It is also worth noting that takings law is notoriously complex and different state to state, and takings challenges are extremely difficult to bring and win. It will likely be many years, or possibly even decades, until it is known how takings issues impact the regulation of shale gas development, if at all.

Constitutions also place real substantive limitations on *how* governments may pass laws, which, in certain cases, can actually support local authority over issues related to shale gas development. Constitutional limitations on laws regarding the use of land provide a good example of this. In general, the U.S. Constitution requires that all laws have some rational relationship to benefiting a public good—a concept known as "substantive due process," which protects against the passage of irrational or arbitrary laws. In the context of zoning and land use law, substantive due process often requires that zoning restrictions accord with a local "comprehensive plan." In other words, zoning controls may be declared unconstitutional where they are not locally tailored or geared toward a common land use plan, such as when a zoning change unfairly benefits select property owners over others or does not adequately take existing local conditions into account.[74]

In most cases, this brand of substantive due process law rarely applies outside of the municipal context, but one recent case from Pennsylvania, *Robinson Township v. Commonwealth*, demonstrates how important it may be in the debate over how shale gas development must be regulated.[75] In *Robinson*, several municipalities challenged portions of newly passed amendments to the Pennsylvania Oil and Gas Act requiring all local zoning of oil and gas activities conform to highly permissive state standards. In practice, these standards would have opened up most of the state to hydraulic fracturing by mandating that oil and gas extraction be permitted as-of-right in all zones (including residential zones) and also by substantially restricting or eliminating local power to establish setbacks and enforce local laws on noise, lighting, or hours of operation. This de facto statewide "zoning ordinance" was declared unconstitutional by both the Pennsylvania appellate-level Commonwealth Court and the state Supreme Court. The Commonwealth Court found that the new act's dramatic and uniform statewide modification of the regulation of an industrial use could not possibly comport with a legitimate comprehensive plan for each and all of Pennsylvania's municipalities, and thus it violated substantive due process principles. The Supreme Court affirmed, but on the separate grounds that the statewide land use provisions violated the state's obligations under the Pennsylvania Constitution's Environmental Rights Amendment.

Without getting into the detailed legal reasoning of these opinions, much of which is specific to Pennsylvania and its state Constitution, the *Robinson* case demonstrates an important and broadly applicable point; namely, that state power to regulate shale gas development may be limited by other constitutional rights, such as the right to the well-considered regulation of land or a right to clean air and water. Additionally, where these rights mandate site-specific consideration of the impacts of state regulatory policy, greater local authority over or input into the regulatory process may be required. Because *Robinson* is the first major constitutional case addressing these types of issues, it may prove to be influential in other shale gas producing states across the country.

SUMMARY

This chapter has tried to illustrate the extent to which governments (federal, state, and local) exercise control over gas development activities. The legal issues are hugely complex and, in many instances, contentious. The role of the federal government in regulating shale gas development is relatively limited. That being said, the federal government does regulate shale gas development in some important ways—for example, monitoring water and air quality through various acts passed by Congress. However, because of a number of exemptions from many of the major federal environmental laws and loopholes inherent in current law, federal authority is restricted.

State governments have the primary responsibility for regulating oil and gas development, and state laws provide nearly unfettered authority for regulation. State regulations, for example, cover testing for gas, well spacing, setbacks, casing and cementing to protect underground water supplies, on-site storage of hydraulic fracturing and drilling fluids and fracking waste, disposal of waste, chemical disclosure requirements, testing water supplies, reporting and remediation of spills, and issues related to property rights surrounding shale gas development. Not surprisingly, rules and regulations vary greatly from state to state.

While the process of shale gas development is often regulated at the state and federal levels, most development (not on federal lands) occurs on land within the jurisdiction of a local government. In fact, regulatory laws pertaining to unconventional gas development, although still evolving, give greater power to state and local governments primarily because of the legal structure in the United States. Local zoning and land use laws are greatly influential in allowing (or not) gas development. Furthermore, many municipalities have laws that apply to broad classes of activities that affect shale gas development, such as air quality ordinances and noise ordinances. However, there is debate at the state level over the extent to which state oil and gas laws override or preempt local laws. Where a local law conflicts with a state statute, the local law is generally preempted.

The legal and regulatory climate is highly political, and oversight over shale gas development is no exception. State and local laws and regulations vary tremendously, and exceptions, usually favoring industry, abound.

NOTES

1. Safe Drinking Water Act (SWDA). http://water.epa.gov/lawsregs/rulesregs/sdwa (accessed June 25, 2014).
2. Underground Injection Control Program. http://water.epa.gov/type/groundwater/uic/regulations.cfm (accessed June 25, 2014).
3. epa.gov/type/groundwater/uic/.../hfdieselfuelsguidance508.pdf (accessed June 25, 2014).
4. 53 Fed Reg. 25447 (June 6, 1988).
5. Energy Policy Act of 2005§ 322, codified at 42 U.S.C. § 300h(d)(1)(B)(ii).
6. Clean Water Act (CAA). 33 U.S.C. §1251 et seq. (1972). http://www2.epa.gov/laws-regulations/summary-clean-water-act (accessed June 25, 2014).
7. 33 U.S.C. § 1362(24).

8. National Pollutant Discharge Elimination System Program. 33 U.S.C. § 1362(6)(B). http://www.epa.gov/compliance/monitoring/programs/cwa/npdes.html (accessed June 25, 2014).

9. Clean Air Act. 42 U.S.C. §§ 7408, 7409, 7412. http://www2.epa.gov/laws-regulations/summary-clean-air-act (accessed June 25, 2014).

10. National Emission Standards for Hazardous Air Pollutants. U.S.C. § 7412. http://www.epa.gov/compliance/monitoring/programs/caa/neshaps.html42 (accessed June 25, 2014).

11. 42 U.S.C. § 7412(n)(4)(A), (B).

12. Clean Air Act: EPA Should Improve the Management of Its Air Toxics Program (GAO-06-669). Published June 23, 2006; publicly released: July 26, 2006. http://www.gao.gov/products/GAO-06-669 (accessed June 25, 2014).

13. Prevention of Significant Deterioration Program. 42 U.S.C. § 7469(1). http://www.epa.gov/NSR/psd.html (accessed June 25, 2014).

14. Nonattainment Area Program. 42 U.S.C. § 7501(c)(5). http://www.epa.gov/air/urbanair/sipstatus/nonattainment.html (accessed June 25, 2014).

15. U.S. EPA. Office of Research and Development. Draft Plan to Study the Potential Impacts of Hydraulic Fracturing on Drinking Water Resources. http://yosemite.epa.gov/sab/sabproduct.nsf/0/D3483AB445AE61418525775900603E79/$File/Draft+Plan+to+Study+the+Potential+Impacts+of+Hydraulic+Fracturing+on+Drinking+Water+Resources-February+2011.pdf. (accessed June 23, 2014).

16. Resource Conservation and Recovery Act. 42 U.S.C. §§ 6901 et seq. http://www2.epa.gov/laws-regulations/summary-resource-conservation-and-recovery-act (accessed June 25, 2014).

17. 42 U.S.C. § 6921(b)(2); 53 Fed. Reg. 25,445 (1988).

18. 42 U.S.C. § 6973.

19. Exemption of Oil and Gas Exploration and Production Wastes from Federal Hazardous Waste Regulations (2002), http://www.epa.gov/osw/nonhaz/industrial/special/oil/oil-gas.pdf. (accessed June 23, 2014).

20. The Comprehensive Environmental Response, Compensation, and Liability Act (CERCLA). 42 U.S.C. §§ 9601 et seq. http://www.epa.gov/superfund/policy/cercla.htm (accessed June 25, 2014).

21. McKay DL, Cook JE, Kennedy RG, et al. RCRA's oil field wastes exemption and CERCLA's petroleum exclusion: are they justified? *Journal of Energy, Natural Resources & Environmental Law* 15 (1995): 41.

22. The Emergency Planning and Community Right-to-Know Act. 42 U.S.C. §§ 11001 et seq. http://www.epa.gov/oecaagct/lcra.html (accessed June 25, 2014).

23. Toxics Release Inventory. 42 U.S.C. § 11023(j). http://www2.epa.gov/toxics-release-inventory-tri-program (accessed June 25, 2014).

24. U.S. EPA. Standard Industrial Classification (SIC) Codes in TRI Reporting, http://www.epa.gov/tri/report/siccode.htm#original_industries (accessed June 25, 2014).

25. 42 U.S.C. §§ 11004, 11021–23.

26. National Environmental Policy Act. 42 U.S.C. §§ 4321 et seq. http://www.epa.gov/compliance/nepa (accessed June 25, 2014).

27. 42 U.S.C. § 15924.

28. Effluent Limitation Guidelines. http://water.epa.gov/scitech/wastetech/guide (accessed June 25, 2014).

29. U.S. EPA. *Permitting Guidance for Oil and Gas Hydraulic Fracturing Activities Using Diesel Fuels: Underground Injection Control Program Guidance #84*. 2014. http://water.epa.gov/type/groundwater/uic/class2/hydraulicfracturing/upload/epa816r14001.pdf (accessed June 23, 2014).

30. Ibid.

31. New Source Performance Standards. 40 CFR Part 63. http://www.epa.gov/compliance/monitoring/programs/caa/newsource.html (accessed June 25, 2014).

32. 77 Fed.Reg. 49490 (Aug. 16, 2012), codified at 40 CFR Parts 60 and 63.

33. U.S. EPA Issues Final Air Rules for the Oil and Natural Gas Industry. http://www.epa.gov/airquality/oilandgas/whitepapers.html (accessed June 25, 2014).

34. Vann A. *Energy Projects on Federal Lands: Leasing and Authorization* (Report R40175). Congressional Research Service, 2012.

35. Gorte RW, Vincent CH, Hanson LA, et al. *Federal Land Ownership: Overview and Data*. Congressional Research Service, February 8, 2012.

36. 78 Fed.Reg. 31636 (May 24, 2013).

37. Toxic Substances Control Act. 15 U.S.C. §§ 2601 et seq. http://www.epa.gov/oecaagct/lsca.html (accessed June 25, 2014).

38. See http:// http://www.epa.gov/oppt/existingchemicals/pubs/tscainventory (accessed June 23, 2014).

39. Endangered Species Act of 1973. http://www.fws.gov/laws/lawsdigest/esact.html (accessed June 25, 2014).

40. U.S. Dept. of Labor, Occupational Safety and Health Administration. Oil and Gas Well Drilling, Servicing and Storage Standards. http://www.osha.gov/SLTC/oilgaswelldrilling/standards.html (accessed June 23, 2014).

41. New York Department of Environmental Conservation. *Guidelines for Seismic Testing on DEC Administered State Land*. http://www.dec.ny.gov/docs/lands_forests_pdf/sfseismic.pdf (accessed June 25, 2014).

42. Wiseman H, Gradijan F. *Regulation of Shale Gas Development, Including Hydraulic Fracturing*. Austin: The Energy Institute, The University of Texas, October 31, 2011.

43. Richardson N, Gottlieb M, Krupnick A, et al. *The State of Shale Gas Regulation*. Resources for the Future, June 2013. http://www.rff.org/rff/documents/RFF-Rpt-StateofStateRegs_Report.pdf (accessed June 23, 2014).

44. Ibid.

45. Riven G. Science is scant as fracking regulators set drilling buffers. *North Carolina Health News*, February 6, 2014. http://www.northcarolinahealthnews.org/2014/02/06/science-is-scant-as-fracking-regulators-set-drilling-buffers (accessed June 23, 2014).

46. Jackson RB, Vengosh A, Darrah TH, et al. Increased stray gas abundance in a subset of drinking water wells near Marcellus shale gas extraction. *Proceedings of the National Academy of Sciences, PNAS Early Edition* (June 9, 2013). http://www.pnas.org/content/early/2013/06/19/1221635110.full.pdf+html (accessed June 25, 2014).

47. Richardson et al., *The State of Shale Gas Regulation*, op. cit.

48. Wiseman and Gradijan, *Regulation of Shale Gas Development*, op. cit.

49. North Dakota Admin. Code 43-02-05-07 (2014).

50. Wiseman and Gradijan, *Regulation of Shale Gas Development*, op. cit.

51. Richardson et al., *The State of Shale Gas Regulation*, op. cit.

52. Hammer R, Van Briesen J. *In Fracking's Wake: New Rules Are Needed to Protect Our Health and Environment from Contaminated Wastewater*. http://www.nrdc.org/energy/files/fracking-wastewater-fullreport.pdf (accessed June 25, 2014).

53. Ellsworth WL. *Abstract: Are Seismicity Rate Changes in the Midcontinent Natural or Manmade?* U.S. Geological Survey, 2012. http://www.fossil.energy.gov/programs/gasregulation/authorizations/Orders_Issued_2012/65._Are_Seismicity_Rate_or_Manmade_.pdf (accessed June 23, 2014).

54. 58 Pa. Cons. Stat. § 3218(c), (c.1), (d) (2014).

55. Wiseman and Gradijan, *Regulation of Shale Gas Development*, op. cit.

56. New York Environmental Conservation Law § 23-0901.

57. McFeeley M. *State Hydraulic Fracturing Disclosure Rules and Enforcement: A Comparison*. Natural Resources Defense Council. http://www.nrdc.org/energy/files/Fracking-Disclosures-IB.pdf (accessed June 25, 2014).

58. Ziegler EH, Rathkopf AH, Rathkopf DA, Rathkopf CA. *Rathkopf's The Law of Zoning and Planning*. New York: Clark Boardman Callaghan, 1993.

59. Pennsylvania Oil and Gas Act (Act 223). www.legis.state.pa.us/WU01/LI/LI/US/HTM/2012/0/0013.HTM (accessed June 25, 2014).

60. Code of Ordinances. Town of Flower Mound, TX. Ch. 34, Art VII.

61. Ohio Constitution. http://www.lsc.state.oh.us/membersonly/128municipalhomerule.pdf (accessed June 25, 2014).

62. Montana Constitution. Art XI, §6.

63. N.Y. Oil, Gas and Solution Mining Law Declaration of Policy (§23-0301). www.dec.ny.gov/energy/26498.html (accessed June 25, 2014).

64. Michigan Comp. Laws §125.3205(b)(2).

65. North Dakota Cent. Code Ann. §38-08-05.

66. *Huntley & Huntley, Inc. v. Borough Council of Borough of Oakmont*, 600 Pa. 207, 225 (Pa. 2009); *Bd. of Cnty. Comm'rs, La Plata Cnty. v. Bowen/Edwards Associates, Inc.*, 830 P.2d 1045, 1057-59 (Colo. 1992).

67. *Northeast. Natural Energy, LLC v. City of Morgantown*, No. 11-C-411, slip op. 6285 (Cir. Ct. Monongalia Cty. August 12, 2011.

68. *Wallach v. Town of Dryden*. Case No. 130, 2014 WL 2921399, *25-7 (2014); *Cooperstown Holstein Corporation v. Town of Middlefield*. Case No. 131, 2014 WL 2921399, *25-7 (2014) (quoting *Gernatt Asphalt Products, Inc. v. Town of Sardinia*. 87 N.Y.2d 668, 684 (1996)).

69. *Voss v. Lundvall Bros., Inc.*, 830 P.2d 1061, 1068 (Colo. 1992).

70. Texas Constitution. Article XI, § 5; *Unger v. State*, 629 S.W.2d 811, 812 (Tex. App. 1982).

71. *Beaver Gasoline Co. v. Zoning Hearing Bd. of Borough of Osborne*, 445 Pa. 571 (1971).

72. California Constitution. Article I, § 19; Michigan Constitution Article. X, § 2; Oklahoma Constitution Article II, § 24.

73. Booher MT. Takings Clause Takes Center Stage in NY Fracking Dispute, Law 360. http://www.law360.com/articles/495670/takings-clause-takes-center-stage-in-ny-fracking-dispute (accessed May 20, 2014).

74. Ziegler et al., *Rathkopf's The Law of Zoning and Planning*, op. cit.

75. *Robinson Twp. v. Commonwealth*, 52 A.3d 463, 482-85 (Pa. Commw. Ct. 2012) affirmed in part, revised in part sub nom. Robinson Twp., *Washington Cnty. v. Commonwealth*, 83 A.3d 901 (Pa. 2013).

Chapter 10

The Ethics of Shale Gas Development: Values, Evidence, and Policy

Jake Hays and Inmaculada de Melo-Martín

INTRODUCTION

The recent boom in natural gas production has resulted in one of the most public environmental policy debates of modern time. Domestic production of natural gas from shale formations is reshaping the U.S. energy economy.[1] Novel constellations of drilling and well stimulation techniques have enabled an economically feasible means of obtaining this resource from new reserves throughout North America. Natural gas produced from shale formations has quickly come to account for a significant portion of overall natural gas production in the United States (roughly 40% in 2012, up from just 2% in 2000).[2] Other countries with significant shale gas reserves are currently looking to exploit this resource as well, intending to capitalize on the perceived economic benefits of a resource that is domestic and relatively bountiful.

With few exceptions, when evaluating technologies in general and energy-related technologies in particular, concerns about ethical values are usually limited to assessing the risks and benefits of technological development and implementation. This has not been different in the case of shale gas development. In general, the main or only ethical concern has focused on the trade-off between the risks and potential benefits that this unconventional energy production technique presents to human health, the environment, energy availability, or economic productivity. In these discussions, economic considerations are typically weighed against environmental and public health concerns. Proponents argue that the economic benefits outweigh the environmental risks, while opponents argue that natural gas extracted from shale formations comes at too great of a cost to the environment and public health.[3]

Discussions on the ethical consequences of shale and tight gas development often focus on attempting to answer scientific questions related to the impact of drilling and well stimulation techniques on the air, water, geology, climate, economy, and public health. For instance, a growing body of research examines the greenhouse gas (GHG) emissions and climate change impacts of shale gas development. This research provides insight into the relative advantages or disadvantages of shale gas by comparing its greenhouse gas footprint to coal, a fossil fuel that natural gas could replace in the electricity generation sector. According to some evidence, closing coal-fired power plants and substituting them with natural gas can create reductions in GHG emissions.[4] However, an increasing body of literature suggests that when life-cycle methane emissions are accounted for, the climate benefits of natural gas are greatly diminished.[5–9]

Regarding the potential for water contamination, published studies have compared methane (the principal component of natural gas) concentrations in private drinking wells in areas with active shale gas development with areas where development is not occurring,[10–12] and other association studies have done the same with contaminants such as heavy metals.[13] Modeling studies have examined whether pathways would allow for the transport of contaminants from fractured shale to aquifers.[14] Other investigations have focused on failure mechanisms such as well barrier and integrity failure rates[15] to quantify the extent to which shale gas development is a threat to our water resources and potentially to public health.

Other potential environmental impacts have been examined as well. Air quality studies have attempted to measure emissions and atmospheric concentrations of contaminants associated with shale and tight gas development,[16–18] sometimes to assess the relative risks to human health.[19,20] Ecological studies have been conducted on particular species of animals,[21–23] on forests,[24] and on habitat fragmentation[25] and biodiversity.[26,27] Induced seismicity and earthquake activity have also been examined because parts of the development process (e.g., fluid/waste injection) have been linked with seismic activity in various parts of the country.[28,29]

Studies on economic impacts are also often considered when trying to determine the ethical consequences of developing shale gas. For instance, research has been conducted to gauge energy return on investment and well productivity in particular regions,[30,31] impacts on other industries such as agriculture,[32] international markets for natural gas,[33] local and state economies,[34] as well as employment and income.[35,36] Both environmental and economic research are sometimes compared to existing energy development options, such as other fossil fuels and renewables.[37]

Concerns about the uncertainty regarding risks and benefits of drilling and well stimulation complicate discussions on ethical concerns. Ethical values, however, play an essential role not just in the evaluation of risks and potential benefits but also in the production of scientific and technological knowledge as well as in science-based policy. Values may shape research agendas, influence the interpretation of the evidence, and direct the way science is used to promote particular policies.[38] For shale gas development, the interpretation and use of scientific evidence is tightly bound with a variety of value judgments. Acknowledgment and critical evaluation of such judgments is always important in scientific discussions, but it is particularly so when dealing with potentially environmentally contentious issues such as energy production.

This chapter provides an overview of the various ways in which ethical judgments play a role in the development and implementation of policy decision making as related to shale gas. First, we explore the role of values in scientific reasoning. Second, we consider the ways in which ethical value judgments shape preferences for particular policies regarding shale gas development. We then discuss specific value considerations that should be taken into account alongside traditional risk assessments for shale gas development. We conclude by making the case for the importance of attending to value considerations in conversations about energy and environmental policy.

VALUES AND SCIENTIFIC REASONING

Although often unrecognized, value judgments play legitimate roles throughout the research process[39] as well as in policy-making decisions.[40] Scientists must make choices on a variety of aspects—such as the hypothesis to be tested, the boundaries of what and who are included and excluded from the analysis, the experimental comparators, the endpoints thought to be relevant, the time frame for observations, the methodology to use, and the interpretation of the evidence. This is also the case for research related to shale gas. In this section, we briefly discuss how values play a role in scientific methodology and the interpretation of evidence in relation to shale gas.

Methodology and Framing: The Case of Global Warming Potential Time Horizons

Methodological choices in science are not only determined by the scientific evidence but are also influenced by a variety of value considerations.

Research on methane emissions and the climate impact of shale gas provides a useful illustration of this point. As previously mentioned, there is a growing body of science on GHG emissions that might occur during a number of stages of shale gas production, transmission, and distribution. This research aims to shed light on shale gas development's contribution to climate change. Methane (CH_4) is a more potent GHG than CO_2. The most recent estimates from the Intergovernmental Panel on Climate Change (IPCC) indicate that CH_4 is approximately 86 times as potent as CO_2 over a 20-year time frame and approximately 34 times as potent over a 100-year time frame.[41] As the primary component of natural gas, CH_4 is leaked and vented into the atmosphere at many stages of production and transmission.[42–44] Significant research efforts are underway,[5,45] yet there is still no scientific consensus on the relative life cycle GHG contribution of shale gas compared with coal.[42,46,47]

The appropriate time horizon for comparing the global warming potential (GWP) of methane (CH_4) has been the subject of some debate. GWPs account for different GHG attributes, relative to carbon dioxide (CO_2), such as the capacity to absorb heat and the lifetime of the compound in the atmosphere. A 100-year time horizon often tends to be used[48] by scientists because CO_2 is the primary GHG of concern due to both its quantity and its longevity in the atmosphere. Using this timeframe allows commensurability across all greenhouse pollutants, and so it seems to make sense to look at all the other GHGs (e.g., CH_4, ozone [O_3], black carbon [BC], nitrous oxide [N_2O], etc.) with reference to CO_2. Furthermore, the societal effects of climate change are projected to be more significant farther into the future, which supports the rationale for choosing a longer (e.g., 100-year) time frame.[49]

However, there are also sound reasons for using a 20-year time horizon when presenting scientific data on climate change.[50] Strong evidence has suggested that warming of the Earth to 1.8° C above the 1890–1910 baseline may result in a significant and rapid increase in the release of methane from the arctic due to the thawing of permafrost.[51] Consensus climate science indicates that this 1.8°C threshold may be reached within the next 30 years unless action is taken to reduce the emissions of methane and other short-lived GHGs that are rapidly warming the earth.[52,53] Crossing such thresholds could bring a number of disastrous economic, environmental, and health effects, the costs of which will be paid for by vulnerable populations in developing countries who will be burdened by severe weather events, poorer health, and lower agricultural production.[51] This rationale is based on the notion that it is imperative to reduce methane emissions over the next 30 years to avoid these temperature thresholds.

There are good reasons, however, for using both 20-year and 100-year time horizons, and, in our opinion, research should account for both time frames to arrive at meaningful scientific and policy conclusions. Although the majority of scientific studies use the 100-year time horizon, this choice relies, in part, on nonepistemic value judgments.[41,54] Implicit in a time horizon of 20, 100, or even 500 years are value judgments about the appropriate action time for addressing climate change. Additional values may include relative concerns for posterity (e.g., populations existing 50 years in the future vs. 200 years), short- versus long-term risk aversion, and environmental conservation and preservation. Each of these value judgments is legitimate and should be recognized as relevant to the interpretation of evidence and policy determinations.

Interpreting the Evidence: The Case of Methane Contamination of Groundwater

Values can also play a role in the interpretation of scientific results on the relative safety of shale gas development in a number of ways. Research on methane contamination of groundwater offers one example of how the interpretation of evidence relies on values as much as science. Evidence from Pennsylvania suggests that methane concentrations in drinking water wells are positively correlated with distance from active shale gas development; that is, higher concentrations of methane were observed the closer an aquifer was to active development.[10,12] However, methane can occur naturally in aquifers, and some conflicting scientific opinion has emerged as to whether this methane contamination is caused or exacerbated by shale gas development.[55–57] Although more research is needed, particularly across geologically diverse shale formations, positions have been taken based on different interpretations of the available evidence.

Scientists tend to agree that methane contamination of drinking water has occurred in Pennsylvania, but they differ in their perceptions of how shale gas development may ascribed to this contamination.[10,12,55–58] Similar to the research on climate change, studies that attempt to observe this potential association involve value judgments. For studies on groundwater contamination, values play a role in ascribing particular importance to water resources as well as to the importance of funding or conducting studies that address these concerns.

Judgments about the importance of water resources, environmental conservation/preservation, or trust in the institutions that manage risks can lead scientists to interpret the evidence on methane contamination as supporting or not a causal connection between the presence of methane and shale gas.[58] Such judgments can lead some scientists to focus on the limitations of the studies, suggesting that with an absence of baseline data there is no evidence of causation. In these cases, they may argue that without such evidence, no meaningful connections can be made between shale gas development and water contamination. They may look past suggestions for more research and focus instead on the absence of present information. Concerns for water contamination or environmental conservation, however, may lead other scientists to take the presence of methane as relevant, even if the causality of the effects remains open, and call for more research on this possible association and the implications of the safety of shale gas development.

VALUES AND POLICY

The common belief that empirical data and not values should inform policy is also present in debates about shale gas. Given this belief, it should come as no surprise that nonepistemic values are often ignored in discussions about science-based policies related to this new energy source. Such discussions are thus construed solely as pertaining to empirical questions. Yet, as in the case of scientific reasoning, judgments about whether existent evidence may or may not support a particular policy are also grounded on ethical values.[59,60]

When science is used to guide policy, questions arise about what constitutes sound scientific evidence, how much is required to draw appropriate conclusions, and which studies are sufficiently reliable. In the shale gas debate, both sides believe that the scientific evidence is on their side and have accused the other of cherry-picking information, misinterpreting evidence, or, worse, purposefully manipulating data. Despite the limited data and subsequent uncertainty about the relative safety of shale gas development, firm positions have been taken on both sides.

Disagreements about Risks and Benefits

Few would argue that shale gas development is without risks to the environment and public health or that it presents no potential benefits to the economy and our energy supply. However, the extent to which these risks may or may not be considered serious or manageable, or whether potential benefits outweigh such risks, largely depends on particular value judgments about how much significance should be given to particular risks, how much risk is acceptable, and how competing risks should be weighed. Specific ethical value judgments about how safe is "safe enough" and how much trust to place on the management of risks lead proponents and opponents (i.e., those more or less sympathetic of this new energy technology) to different conclusions about the sufficiency of the available evidence and whether such evidence supports or undermines the continued development of this resource.

For instance, one disagreement involves the weight that should be given to worst-case scenarios related to environmental degradation or harms to public health. Proponents tend to judge that the benefits society might receive from shale gas development, such as economic growth, energy sufficiency, and jobs, are sufficient to outweigh the inherent risks.[3] It is not that they do not acknowledge such risks at all, but they deem them insignificant or manageable through adequate regulation and technological improvements.[61,62] Opponents, on the other hand, judge these potential benefits as overblown[63] and see risks to the environment and public health as deserving more consideration than proponents of this resource usually give to them.[64] They are also less convinced about the ability of regulations to adequately manage risks because of the limited resources and capabilities of regulatory agencies.[65] Moreover, opponents of shale gas development judge that regulations are unlikely to make the process safer primarily because of the inherent risks of the extraction process. Also, there is a tendency for companies to fail to abide by regulatory restrictions. The opponents call attention to the difference between *best* practices and *actual* practices.

Proponents seem to view shale gas development as a generally safe activity that can be risky at times. Because proponents may not see shale gas development as particularly risky, evidence of harmful environmental and public health outcomes are considered avoidable through best practices. While opponents point to adverse outcomes as reasons to stop development, proponents interpret these outcomes as manageable. Focus is then placed on technological solutions to engineering problems and mitigation techniques. The question becomes not *whether* certain industrial problems can be remedied, but *how*.

Comparisons: Better versus Not Good Enough

Proponents and opponents also tend to disagree on the appropriate comparisons to other forms of energy production and use. Because both coal and natural gas are used for electricity generation and because natural gas has generally been thought to be a suitable energy source to displace coal, proponents of shale gas development usually compare it to coal production.[61] Natural gas is cleaner than coal from an end-use standpoint because it emits less CO_2 into the atmosphere during combustion. Evidence also suggests that converting coal power plants to natural gas will eventually lessen the climatic effects of emissions.[66] For some proponents, transitioning away from coal should be our first priority, and they see natural gas as providing such an opportunity. They tend to view shale gas as the lesser of two evils, characterizing natural gas as a "bridge fuel" to a cleaner energy future.

For some opponents, the appropriate comparison is not with coal but with renewable energy sources. They argue that relying on shale gas detracts from moving forward with cleaner, more sustainable forms of energy. They believe that this is a path in the wrong direction because it prolongs reliance on fossil fuels rather than serving as a bridge to renewable energy.[67] Indeed, opponents often maintain that even if evidence shows that natural gas is better than coal from an environmental point of view, it may not be "good enough" compared with more sustainable forms of energy production.[68]

Choices about the appropriate source of comparison are grounded on value judgments; some of them concern pragmatic considerations. Although the economic and technical feasibility of converting all-purpose energy infrastructures has been modeled,[68,69] proponents and opponents judge differently the feasibility of actual implementation due to political will and zoning restrictions. Similarly relevant is the relative weight given to the interest of future generations. In some instances, coal may serve as a more appropriate comparison reference for concerns about present generations and the benefits provided to them by short-term economic growth. Renewable sources of energy, however, may offer a more adequate point of comparison when paying attention to the interests of future generations.

False Positives and False Negatives

Despite scientific uncertainty, governments need to make decisions about how to proceed with developing shale gas. Local, state, and national legislatures have varied significantly in this regard. For instance, some states, such as Pennsylvania, Texas, and Colorado, have explicitly sanctioned the development of this resource, whereas Vermont has banned the process outright. Others such as New York and Maryland have chosen to postpone development through moratoria until the risks are better understood.

Decisions about whether to allow development and implementation of shale gas not only are grounded on the existent scientific evidence but also reflect a legislature's ethical preferences to avoid false-negative or false-positive errors. A false-positive occurs when a true null hypothesis is rejected, suggesting an effect exists when one does not. For instance, for the null hypothesis "shale gas development has no detrimental effects on the health of populations," a false positive occurs when a determination is made that detrimental effects do exist, when such is not the case. A false-negative occurs when a false null hypothesis fails to be rejected, suggesting a lack of effect when one does exist. Under the same null hypothesis, a false-negative would involve accepting that there are no detrimental effects of shale gas development on the health of populations when such effects exist.

When considering policy options in situations of uncertainty, policymakers must risk either rejecting a true null hypothesis or failing to reject a false null hypothesis. In other words, they risk either failing to develop shale gas when the technology is safe or developing shale gas when doing so is unsafe. Policymakers must determine which of the two risks is preferable when minimizing both is not possible. Minimizing false positives lowers the possibility of restricting a harmless activity. Minimizing false negatives limits the possibility of accepting a harmful activity.

What type of error to minimize is not an easy decision, and problematic outcomes may result from either preference. Minimizing false positives can lead to the underregulation of certain practices, which may result in harmful outcomes (e.g., environmental degradation, adverse health outcomes). On the other hand, minimizing false negatives can lead to overregulation, which may impede positive outcomes (e.g., jobs, economic development).

Preferences for minimizing false negatives or false positives in situations of uncertainty may arise for a variety of reasons. For instance, minimizing false positives may appear to be more consistent with scientific practice because scientists prefer to fail to discover truths rather than accept falsehoods. Yet public policy and science have different goals and thus minimization of false negatives may be more appropriate when developing public policy in situations of uncertainty.[70] The preference between minimizing false negatives or false positives involves value judgments about how competing interests should be balanced in light of uncertainty or disagreements about the available evidence. In some situations, there may be ethical reasons to support a preference to minimize false negatives.[71] In others, there may be reasons to support a preference to minimize false positives.[72]

A variety of value judgments can ground decisions about whether to minimize false positives or false negatives.[71] For instance, judgments about the relative importance of protecting people from harm versus enhancing people's welfare are the basis for preferences for minimizing false negatives or false positives, respectively. Permitting the development of shale gas technologies may enhance the welfare of some people either directly (e.g., industry jobs or royalty payments for landowners leasing their mineral rights) or indirectly (e.g., increased tax revenues, cheaper natural gas prices for individuals and local, state, and federal governments, as well as other industries using natural gas). If enhancing people's welfare is particularly valued, then legislators are likely to favor shale gas policies that minimize false positives. However, if protection from harm is given more weight, then legislators might be more inclined to favor policies that minimize false negatives.

Similarly, how much importance is given to peoples' opportunities to make autonomous decisions can shape preferences for minimizing false positive or negatives. Judging seriously concerns about the involuntariness of the imposition of risks related to shale gas development and about people's limited access to information relevant to the assessment of risks can ground an interest in policies that minimize false negatives. Also of relevance to decisions over which of these errors to minimize are value judgments about the feasibility of alternatives to shale gas. On the one hand, if shale gas is seen as a better alternative than coal, then policies that minimize false positives would seem reasonable. On the other hand, if legislators judge that renewable sources of energy would provide us with less harmful and similarly beneficial alternatives to conventional sources of energy, then they are likely to favor policies that will minimize false negatives.

Identifying Values

As previously mentioned, ethical concerns in the shale gas debate tend to revolve around assessment of risks usually understood in focused ways, such as risks to public health or the environment. Nonetheless, determinations about the relative safety of shale gas are insufficient to settle policy debates rationally. Other value considerations are also important when deciding ethically sound public policy regarding shale gas development.

Technological developments not only have implications for the environment and public health but they also affect, sometimes profoundly, people's way of living and thus an important aspect of human well-being that is not reducible to health and safety. Shale gas development must be spatially intense to be economically feasible because natural gas is distributed throughout large geographic formations.[73] The industrial build out requires the construction not only of well pads but of ancillary infrastructure to accompany production. This infrastructure is used for hydrocarbon processing, wastewater disposal, transmission and distribution, and requires the construction of dew point facilities, compressor stations, pipelines, wastewater treatment ponds, and other storage facilities. It also requires thousands of heavy truck trips for transporting water, sand, chemicals, and wastewater before, during, and after the production phase.[74]

Much of the drilling and well stimulation activity is occurring in small, rural communities throughout the United States.[32] Consequently, many communities are undergoing significant change with rapid industrialization and influxes in population. Such changes will affect crime rates, traffic accidents, sexually transmitted diseases, and psychological health concerns, such as anxiety and stress, which are usually taken into account in traditional risk assessments.[75,76] However, other aspects important to people's well-being, such as the supplementation of agricultural economies with industrial development and the implications of this transformation for family relationships, or the effects on the ways that humans experience their local environments will also be affected, and thus they should not be neglected when determining ethically sound public policies.

Similarly, development of shale gas can have transformative effects on the empowerment of local communities and popular participation. In New York State, for instance, a growing number of municipalities have used local zoning ordinances to prohibit oil and gas development on local land. However, the economic feasibility of shale gas is largely dependent on spatially intense industrial build-out across New York's Southern Tier. This effort may be thwarted if particular townships have the legal and political ability to restrict the practice, an ability that can constrain the future of oil and gas development in the state. As a result, billion-dollar corporations have brought lawsuits against small towns in bids to force the town to accept shale gas development. Operators in New York claim that the Oil, Gas and Solution Mining Law supersedes local authority and that only the state can regulate this industry. Although local efforts in New York have so far been successful and local bans have been upheld in intermediary courts, local jurisdictions in other states may not have the power to enact bans and moratoria. Democratic participation and the ways that such participation can be affected by technological development are thus also relevant when evaluating shale gas practices.

Equally important is the impact of this new technological development on people's ability to make choices that have profound effects on their lives. For instance, landowners who do not lease their property for industrial purposes may still be subjected to risks by the drilling activities occurring on a neighbor's land and thus they might be compelled into leasing agreements. Landowners can lose the opportunity to make free and informed decisions about participation in this industrial process in other ways as well. For operations to be economically feasible, natural gas companies need to drill extensively into the shale formation. States like New York have a compulsory integration process known as "forced pooling," where holdout landowners are forced to join gas-leasing agreements with their neighbors if leases have been negotiated for at least 60% of the 640-acre land unit of which they are a part. Thirty-nine states have some kind of compulsory integration laws similar to forced pooling in New York.[77]

Furthermore, because people often do not own the minerals found in their properties, other parties that own these rights are able to drill on private properties without the surface owner's consent. The subsurface property takes precedence over surface rights in what is referred to as a "split estate."[78] Given that mineral rights are the dominant property claim, outside interests are able to subject members in communities to shale gas development against their will and without just compensation.

Finally, ethically sound policy determinations about shale gas development should be attentive to the ways in which such development can reinforce, shape, or transform particular beliefs about what constitutes appropriate levels of energy consumption. Much of the argument for unconventional fossil fuel development is predicated on the assumption that society should continue its current level of energy consumption. Assessments of risk and benefits of this new energy resource are thus affected by this presupposition. Yet calling into question such an assumption is bound to have profound effects on our judgments about the risks and benefits involved in the development of shale gas.

CONCLUDING THOUGHTS: THE IMPORTANCE OF VALUES

Value judgments can and should play a legitimate role in the interpretation of the scientific evidence and in policy decisions. Acknowledging these values is important for several reasons. First, a discussion about values is crucial to moving the debate forward and enabling meaningful communication among researchers. Without this discussion the debate will remain stagnant. Proponents of this energy source, for instance, will offer reasons why developing shale gas is a safe alternative, but because opponents do not share the same value judgments, such reasons will not be recognized as good ones. This situation is unhelpful for scientists, policymakers, and the general public. An honest conversation about values can help illuminate what is really at stake and how value judgments affect interpretations of the empirical evidence and policy choices.

Second, failing to recognize and consider relevant value judgments may make research efforts less productive or even irrelevant. The debate will continue to focus on the need to obtain more scientific evidence. Although less uncertainty can help solve some of the questions at stake in the development of shale gas, it will not resolve the disagreements. For instance, if one judges as important the interests of future generations and reasonably believes a transition to renewable energy from fossil fuels is feasible, then showing that natural gas contributes less to global warming will do little to convince opponents of shale gas. In these instances, researchers may be obtaining adequate information, but for the wrong questions. Having an upfront conversation about values will help facilitate agreements about the research agenda.

Finally, engaging with the underlying values at stake in these debates will help ensure sound public policy. We want our public policies to be grounded on good science, and good science requires careful attention to the evidence, adequate research methodologies, and appropriate research questions. But attention to the value judgments that underlie such research is also essential. If value judgments about how safe is "safe enough" or about whether the interests of future generations should be considered are not acknowledged and critically assessed, it might well be that such judgments will not be rationally supported. Hence, good policymaking will be well served by careful attention to value judgments.

NOTES

1. U.S. Energy Information Administration. *Annual Energy Outlook for 2013.* http://www.eia.gov/forecasts/aeo/pdf/0383(2013).pdf (accessed February 5, 2014).

2. Hughes JD. Energy: A reality check on the shale revolution. *Nature* 494 (2013): 307–308.

3. Howarth RW, Ingraffea A, Engelder T. Natural gas: Should fracking stop? *Nature* 477 (2011): 271–275.

4. Pacala S, Socolow R. Stabilization wedges: Solving the climate problem for the next 50 years with current technologies. *Science* 305 (2004): 968–972.

5. Pétron G, Karion A, Sweeney C, et al. A new look at methane and non-methane hydrocarbon emissions from oil and natural gas operations in the Colorado Denver-Julesburg Basin. *Journal of Geophysical Research: Atmospheres* 119 (2014): 6836–6852.

6. Caulton DR, Shepson PB, Santoro RL, et al. Toward a better understanding and quantification of methane emissions from shale gas development. *Proceedings of the National Academy of Sciences of the United States of America* 111 (2014): 6237–6242.

7. Karion A, Sweeney C, Pétron G, et al. Methane emissions estimate from airborne measurements over a western United States natural gas field. *Geophysical Research Letters* 40 (2013): 4393–4397.

8. Brandt AR, Heath GA, Kort EA, et al. Methane leaks from North American natural gas systems. *Science* 343 (2014): 733–735.

9. Miller SM, Wofsy SC, Michalak AM, et al. Anthropogenic emissions of methane in the United States. *Proceedings of the National Academy of Sciences of the United States of America* 110 (2013): 20018–20022.

10. Jackson RB, Vengosh A, Darrah TH, et al. Increased stray gas abundance in a subset of drinking water wells near Marcellus shale gas extraction. *Proceedings of the National Academy of Sciences of the United States of America* 110 (2013): 11250–11252. doi:10.1073/pnas.1221635110.

11. Warner NR, Kresse TM, Hays PD, et al. Geochemical and isotopic variations in shallow groundwater in areas of the Fayetteville Shale development, north-central Arkansas. *Applied Geochemistry* 35 (2013): 207–220.

12. Osborn SG, Vengosh A, Warner NR, Jackson RB. Methane contamination of drinking water accompanying gas-well drilling and hydraulic fracturing. *Proceedings of the National Academy of Sciences of the United States of America* 108 (2011): 8172–8176. doi:10.1073/pnas.1100682108.

13. Fontenot BE, Hunt LR, Hildenbrand ZL, et al. An evaluation of water quality in private drinking water wells near natural gas extraction sites in the Barnett Shale Formation. *Environmental Science and Technology* 47 (2013): 10032–10040.

14. Myers T. Potential contaminant pathways from hydraulically fractured shale to aquifers. *Ground Water* 50 (2012): 872–882.

15. Davies RJ, Almond S, Ward RS, et al. Oil and gas wells and their integrity: implications for shale and unconventional resource exploitation. *Marine and Petroleum Geology* 56 (2014): 239–254. http://dx.doi.org/10.1016/j.marpetgeo.2014.03.001.

16. Moore CW, Zielinska B, Petron G, Jackson RB. Air impacts of increased natural gas acquisition, processing, and use: a critical review. *Environmental Science and Technology* 48 (2014): 8349–8359. doi:10.1021/es4053472.

17. Pétron G, Frost G, Miller BR, et al. Hydrocarbon emissions characterization in the Colorado Front Range: a pilot study. *Journal of Geophysical Research* 117 (2012): D4. doi:10.1029/2011JD016360.

18. Roy AA, Adams PJ, Robinson AL. Air pollutant emissions from the development, production, and processing of Marcellus Shale natural gas. *Journal of the Air & Waste Management Association* 64 (2014): 19–37.

19. Bunch AG, Perry CS, Abraham L, et al. Evaluation of impact of shale gas operations in the Barnett Shale region on volatile organic compounds in air and potential human health risks. *Science of the Total Environment* 468–469 (2014): 832–842.

20. McKenzie LM, Witter RZ, Newman LS, Adgate JL. Human health risk assessment of air emissions from development of unconventional natural gas resources. *Science of the Total Environment* 424 (2012): 79–87.

21. Weltman-Fahs M, Taylor JM. Hydraulic fracturing and brook trout habitat in the Marcellus Shale region: potential impacts and research needs. *Fisheries* 38 (2013): 4–15.

22. Papoulias DM, Velasco AL. Histopathological analysis of fish from Acorn Fork Creek, Kentucky, exposed to hydraulic fracturing fluid releases. *Southeastern Naturalist* 12, special issue 4 (2013): 92–111.

23. Hamilton LE, Dale BC, Paszkowski CA. Effects of disturbance associated with natural gas extraction on the occurrence of three grassland songbirds. *Avian Conservation and Ecology* 6 (2011): 7.

24. Adams MB. Land application of hydrofracturing fluids damages a deciduous forest stand in West Virginia. *Journal of Environmental Quality* 40 (2011): 1340–1344.

25. Racicot A, Babin-Roussel V, Dauphinais J-F, et al. A framework to predict the impacts of shale gas infrastructures on the forest fragmentation of an agroforest region. *Environmental Management* 53 (2014): 1023–1033.

26. Kiviat E. Risks to biodiversity from hydraulic fracturing for natural gas in the Marcellus and Utica Shales. *Annals of the New York Academy of Science* 1286 (2013): 1–14.

27. Jones IL, Bull JW, Milner-Gulland EJ, et al. Quantifying habitat impacts of natural gas infrastructure to facilitate biodiversity offsetting. *Ecology and Evolution* 4 (2014): 79–90.

28. Elst NJ van der, Savage HM, Keranen KM, Abers GA. Enhanced remote earthquake triggering at fluid-injection sites in the midwestern United States. *Science* 341 (2013): 164–167.

29. Keranen KM, Savage HM, Abers GA, Cochran ES. Potentially induced earthquakes in Oklahoma, USA: links between wastewater injection and the 2011 Mw 5.7 earthquake sequence. *Geology* (2013): G34045.1.

30. Aucott ML, Melillo JM. A preliminary energy return on investment analysis of natural gas from the Marcellus Shale. *Journal of Industrial Ecology* 17 (2013): 668–679.

31. Kaiser MJ. Profitability assessment of Haynesville shale gas wells. *Energy* 38 (2012): 315–330.

32. Finkel ML, Selegean J, Hays J, Kondamudi N. Marcellus Shale drilling's impact on the dairy industry in Pennsylvania: a descriptive report. *New Solutions* 23 (2013): 189–201.

33. Wakamatsu H, Aruga K. The impact of the shale gas revolution on the U.S. and Japanese natural gas markets. *Energy Policy* 62 (2013): 1002–1009.

34. Barth JM. The economic impact of shale gas development on state and local economies: benefits, costs, and uncertainties. *New Solutions* 23 (2013): 85–101.

35. Weber JG. The effects of a natural gas boom on employment and income in Colorado, Texas, and Wyoming. *Energy Economics* 34 (2012): 1580–1588.

36. Kinnaman TC. The economic impact of shale gas extraction: a review of existing studies. *Ecological Economics* 70 (2011): 1243–1249.

37. Jones NF, Pejchar L. Comparing the ecological impacts of wind and oil & gas development: a landscape scale assessment. *PLoS ONE* 8 (2013): e81391.

38. Kitcher P. *Science, Truth, and Democracy*. Oxford: Oxford University Press, 2001.

39. Longino HE. *Science as Social Knowledge: Values and Objectivity in Scientific Inquiry*. Princeton, NJ: Princeton University Press, 1990.

40. Jasanoff S. *States of Knowledge: The Co-production of Science and the Social Order*. London: Routledge, 2006.

41. Intergovernmental Panel on Climate Change. *Climate Change 2013: The Physical Science Basis*. 2013. http://www.ipcc.ch/report/ar5/wg1/#.Uk21dxbhLdk (accessed April 5, 2014).

42. Howarth RW, Santoro R, Ingraffea A. Methane and the greenhouse-gas footprint of natural gas from shale formations. *Climatic Change* 106 (2011): 679–690.

43. Jackson RB, Down A, Phillips NG, et al. Natural gas pipeline leaks across Washington, DC. *Environmental Science and Technology* 48 (2014): 2051–2058.

44. Phillips NG, Ackley R, Crosson ER, et al. Mapping urban pipeline leaks: methane leaks across Boston. *Environmental Pollution* 173 (2013): 1–4.

45. Allen DT, Torres VM, Thomas J, et al. Measurements of methane emissions at natural gas production sites in the United States. *Proceedings of the National Academy of Sciences of the United States of America* 17 (2013): 6237–6242.

46. Cathles LM, Brown L, Taam M, Hunter A. A commentary on "The greenhouse-gas footprint of natural gas in shale formations" by R.W. Howarth, R. Santoro, and A. Ingraffea. *Climatic Change* 113 (2012): 525–535.

47. Howarth RW, Santoro R, Ingraffea A. Venting and leaking of methane from shale gas development: response to Cathles et al. *Climatic Change* 113 (2012): 537–549.

48. U.S. Environmental Protection Agency. *Overview of Greenhouse Gases: Methane Emissions*. http://epa.gov/climatechange/ghgemissions/gases/ch4.html (accessed March 5, 2014).

49. Hausfather Z. *Coal Preferable to Natural Gas from Shale for Climate? Not So Fast . . . And Choice of Time Frame Critical*. The Yale Forum on Climate Change & The Media, 2011. http://www.yaleclimatemediaforum.org/2011/05/coal-preferable-to-natural-gas-or-not (accessed March 15, 2014).

50. Howarth R, Shindell D, Santoro R, et al. *Methane Emissions from Natural Gas Systems*. 2012. http://www.eeb.cornell.edu/howarth/Howarth%20et%20al.%20–%20National%20Climate%20Assessment.pdf (accessed March 5, 2014).

51. Whiteman G, Hope C, Wadhams P. Climate science: vast costs of Arctic change. *Nature* 499 (2013): 401–403.

52. Hansen J, Sato M. Greenhouse gas growth rates. *Proceedings of the National Academy of Sciences of the United States of America* 101 (2004): 16109–16114.

53. Hansen J, Sato M, Kharecha P, Russell G, Lea DW, Siddall M. Climate change and trace gases. *Philosophical Transactions of the Royal Society A: Mathematics, Physics, and Engineering Science* 365 (2007): 1925–1954.

54. Shine KP. The global warming potential—the need for an interdisciplinary retrial. *Climatic Change* 96 (2009): 467–472.

55. Davies RJ. Methane contamination of drinking water caused by hydraulic fracturing remains unproven. *Proceedings of the National Academy of Sciences of the United States of America* 108 (2011): E871–E871.

56. Schon SC. Hydraulic fracturing not responsible for methane migration. *Proceedings of the National Academy of Sciences of the United States of America* 108 (2011): E664–E664.

57. Saba T, Orzechowski M. Lack of data to support a relationship between methane contamination of drinking water wells and hydraulic fracturing. *Proceedings of the National Academy of Sciences of the United States of America* 108 (2011): E663–E663.

58. Jackson RB, Osborn SG, Vengosh A, Warner NR. Reply to Davies: hydraulic fracturing remains a possible mechanism for observed methane contamination of drinking water. *Proceedings of the National Academy of Sciences of the United States of America* 108 (2011): E872–E872.

59. de Melo-Martín I, Intemann K. Interpreting evidence. *Perspectives in Biological Medicine* 55 (2012): 59–70.

60. Elliott KC, Resnik DB. Science, policy, and the transparency of values. *Environmental Health Perspectives* 122 (2014). doi:10.1289/ehp.1408107. http://ehp.niehs.nih.gov/1408107 (accessed November 17, 2014).

61. Hamburg S. *Sustainable Energy for All: Ensuring Health I.* Institute of Medicine of the National Academies, 2012. http://www.iom.edu/Activities/Environment/EnvironmentalHealthRT/2012-APR-30.aspx (accessed April 21, 2014).

62. Groat CG, Grimshaw TW. *Fact-Based Regulation for Environmental Protection in Shale Gas Development.* Austin: The Energy Institute, University of Texas, 2012. http://www.energy.utexas.edu/images/ei_shale_gas_reg_summary1202.pdf (accessed April 21, 2014).

63. Heinberg R. *Snake Oil: How Fracking's False Promise of Plenty Imperils Our Future.* Santa Rosa, CA: Post Carbon Institute, 2013.

64. Sadasivam N. *In Fracking Fight, a Worry about How Best to Measure Health Threats.* ProPublica. April 1, 2014. http://www.propublica.org/article/in-fracking-fight-a-worry-about-how-best-to-measure-health-threats (accessed April 20, 2014).

65. Kusnetz N. *Many PA Gas Wells Go Unreported for Months.* ProPublica. February 3, 2011. http://www.propublica.org/article/many-pa-gas-wells-go-unreported-for-months (accessed March 5, 2014).

66. Lafrancois BA. A lot left over: reducing CO_2 emissions in the United States' electric power sector through the use of natural gas. *Energy Policy* 50 (2012): 428–435.

67. Ingraffea AR. Gangplank to a warm future. *The New York Times*, July 28, 2013. http://www.nytimes.com/2013/07/29/opinion/gangplank-to-a-warm-future.html (accessed March 30, 2014).

68. Jacobson MZ, Howarth RW, Delucchi MA, et al. Examining the feasibility of converting New York State's all-purpose energy infrastructure to one using wind, water, and sunlight. *Energy Policy* 57 (2013): 585–601.

69. Jacobson MZ, Delucchi MA. Providing all global energy with wind, water, and solar power, part I: technologies, energy resources, quantities and areas of infrastructure, and materials. *Energy Policy* 39 (2011): 1154–1169.

70. Shrader-Frechette KS. *Risk and Rationality: Philosophical Foundations for Populist Reforms*. Berkeley: California University Press, 1991.

71. de Melo-Martín I, Hays J, Finkel ML. The role of ethics in shale gas policies. *Science of the Total Environment* 470–471 (2014): 1114–1119.

72. Department for Environment, Food and Rural Affairs. *An Assessment of Key Evidence about Neonicotinoids and Bees*. https://www.gov.uk/government/uploads/system/uploads/attachment_data/file/221052/pb13937-neonicotinoid-bees-20130326.pdf (accessed March 5, 2014).

73. Law A, Hays J. Insights on unconventional natural gas development from shale: an interview with Anthony R. Ingraffea. *New Solutions* 23 (2013): 203–208.

74. New York State Department of Environmental Conservation. *Revised Draft SGEIS on the Oil, Gas and Solution Mining Regulatory Program*. 2011. http://www.dec.ny.gov/energy/75370.html (accessed April 21, 2014).

75. Witter RZ, McKenzie L, Stinson KE, et al. The use of health impact assessment for a community undergoing natural gas development. *American Journal of Public Health* 103 (2013): 1002–1010.

76. Garfield County, Colorado. *Battlement Mesa Health Impact Assessment*. 2011. http://www.garfield-county.com/environmental-health/battlement-mesa-health-impact-assessment-ehms.aspx (accessed April 18, 2014).

77. Baca MC. *Forced Pooling: When Landowners Can't Say No to Drilling*. April 10, 2011. ProPublica. http://www.propublica.org/article/forced-pooling-when-landowners-cant-say-no-to-drilling (accessed March 7, 2014).

78. Bureau of Land Management. *Split Estate*. 2013. http://www.blm.gov/wo/st/en/prog/energy/oil_and_gas/best_management_practices/split_estate.html (accessed April 5, 2014).

Chapter 11

An Industry Perspective on the Benefits and Potential Community Health Risks Related to Unconventional Resource Development

Dennis J. Devlin

"FRACKING"

The United States sits atop tremendous natural resources, including onshore natural gas and oil deposits that are locked in shale and tight geologic formations deep underground. "Unconventional resources" is the collective term used to describe the shale gas and oil, tight gas and oil, and coal bed methane.[1] We have been aware of their existence for decades but until recently did not have commercially viable technology to access and produce them economically. The process of hydraulic fracturing—injecting a high-pressure solution primarily of water and sand mixed with small concentrations of chemicals to fracture rock—had been used in a limited way to produce unconventional resources since the 1940s. Horizontal, or directional, drilling is another technique that had been in use for decades, consisting of drilling vertically to a point and then turning to run horizontally for greater access to resources layered in formations. The breakthrough came in the early 2000s when horizontal drilling was combined with hydraulic fracturing in an economically viable way, unleashing the surge in natural gas and oil production that the United States has experienced since. Unconventional resource development (URD) refers to the entire process, but due to extensive use by the mass media, the term "fracking" is now solidly grounded in the American vernacular and is used to refer to all or part of the URD process. Although this book's focus is primarily natural gas, from an industry perspective, it is important to recognize the significance of unconventional gas and oil, with both being produced by hydraulic fracturing.

RECOGNIZING BENEFITS

The world faces multiple challenges involving energy, economic development, and environmental protection. The oil and gas industry strives to meet these challenges by providing safe, reliable, and affordable supplies of energy in an environmentally responsible manner that helps sustain and improve living standards for populations in the United States and

Figure 11.1
Projection of global energy supplies: 2010–2040.

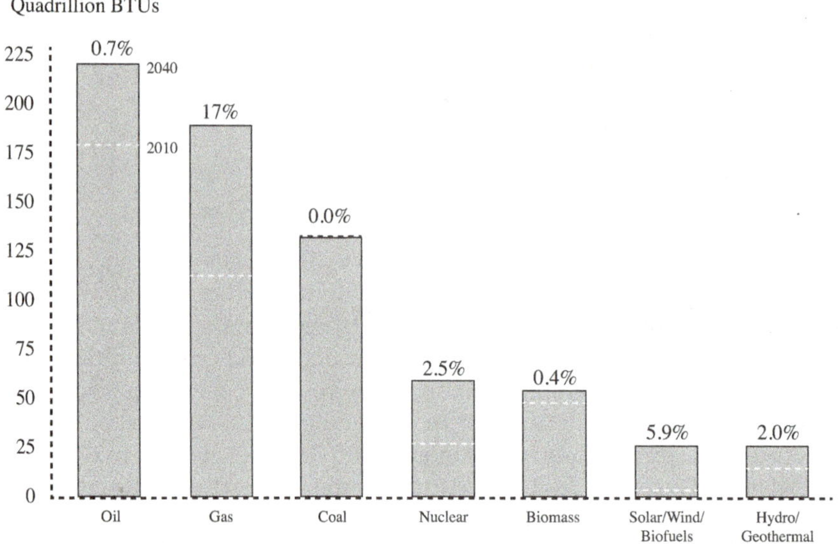

Source: ExxonMobil Corporation.[2]
Note: Each bar represents global energy demand by type, in quadrillion BTUs per year, along with projected average annual growth rate (in percent). The dashed lines represent global demand in 2010, and the full bar represents global demand in 2040.

worldwide. ExxonMobil annually produces an "Outlook for Energy" that looks at trends in the types of energy the world will need based on how that energy is going to be used, as well as the availability of affordable and reliable supplies.[2] Figure 11.1 shows the projection of the energy the world will use in 2040 compared with 2010. Natural gas is expected to grow at an average annual rate of 1.7% with overall demand up about 65% by 2040, whereas coal demand is expected to increase over the medium term before ending about where it is today. The International Energy Agency also projects significant growth for natural gas over the next two decades but at a slightly slower average annual rate of 1.6%, and with coal demand increasing at an average annual rate of 0.7%.[3] This increase in demand for natural gas needs to be matched by an increase in supply but balanced with environmentally sound production. That includes the need to reduce carbon emissions, particularly from power generation.

URD is providing substantial benefits for the United States in terms of economic growth, energy security, and pollution and greenhouse gas emissions reduction, which are discussed below.

Economic Growth

McKinsey & Company estimated in 2013 that shale energy could add 2% to 4% ($380 billion–$690 billion) to annual gross domestic product in the United States by 2020.[4]

- Governments experience increased revenue from a variety of sources related to URD, including severance taxes distributed by the state government, local property taxes and sales taxes, and direct payment from oil and gas companies. URD was responsible for $63 billion in revenue to all levels of government in 2012, and it is estimated that it will exceed $1.4 trillion for the period 2012 through 2025.[5]
- URD supported 1.7 million jobs in the United States—direct and indirect—in 2012, and this number is projected to rise to nearly 3 million by 2020.[6] These jobs are, and will continue to be, an important source of high-wage employment for workers both with and without college degrees, generating economic activity in parts of the United States that have seen limited investment in recent decades. Furthermore, a significant portion of this economic activity is also seen in nonproducing states due to goods and services critical to oil and gas production.
- According to the U.S. Bureau of Labor Statistics, in January 2013 the total U.S. nonfarm payroll employment was 2.3% below the January 2008 level, while new oil and gas jobs increased employment in the energy industry by more than 26% over this period.[7]

- Natural gas is used both to generate electricity and as a raw material in many products, including fertilizers, petrochemicals, fabrics, pharmaceuticals, and plastics. Manufacturers in the United States have predicted $11.6 billion in savings by 2025 due to lower feedstock and energy costs.[8]
- The American Chemistry Council announced in July 2014 that potential U.S. chemical industry investment linked to natural gas and natural gas liquids from shale formations reached 148 projects valued at $100.2 billion.[9] These projects—new factories, expansions, and process changes to increase capacity—could lead to $81 billion per year in new chemical industry output. More than half of the investment is by firms based outside the United States. These companies appear to have concluded that there will be a stable, sustainable supply and cost advantages in the United States with natural gas as a feedstock and energy source. It is estimated that this new investment will help turn a $3 billion trade deficit for chemicals in the United States into a $30 billion surplus in 5 years.[10]

Energy Supply and Security

- More than 60 million U.S. homes are currently heated with natural gas. Furthermore, natural gas is used in many other ways, including heating buildings, heating water, cooking, drying clothes, lighting, and for many industrial purposes. Unconventional gas production grew 51% annually between 2007 and 2012, helping to lower the wellhead price of natural gas by two-thirds.[11]
- In addition to the surging natural gas production from shale regions, there has been an equally significant increase in the production of oil and natural gas liquids. The U.S. Energy Information Administration (EIA) reports that between 2008 and 2014 U.S. oil production jumped nearly 70% from 5 million to 8.5 million barrels per day (bbl/d). It is expected to average 9.3 million bbl/d in 2015, which would be the highest annual average level of oil production since 1972. Natural gas liquids production is expected to increase from an average of 2.6 million bbl/d in 2013 to 3.1 million bbl/d in 2015.[12]
- The growth in domestic production has contributed to a significant decline in petroleum imports. The share of total U.S. petroleum and other liquids consumption met by net imports fell from 60% in 2005 to an average of 33% in 2013. EIA expects the net import share to decline to 22% in 2015, which would be the lowest level since 1970.[13]
- The United States has overtaken Russia as the world's largest natural gas producer and in 2014 surpassed Saudi Arabia as the world's largest oil producer.[14]

- As with natural gas, rising oil production from URD is transforming energy markets and upending traditional notions of American energy scarcity. In December 2012, President Obama highlighted the significant geopolitical nature of these benefits by saying that the United States is going to be *a net exporter of energy* because of new technologies. During his first term in office, the United States reduced its dependence on foreign oil each year; the figure is now under 50%.[15]

Pollution Reduction

- Natural gas will play an important role for a cleaner energy mix around the world as abundant and inexpensive supplies lead utilities to switch from coal to gas in generating electricity. Emissions of natural gas have significantly lower levels of contaminants, such as mercury, sulfur and nitrogen oxides, and particulates.
- The EIA reports that natural gas emits 44% less carbon dioxide (per million BTU) than coal.[16] It also reports that energy-related carbon dioxide emissions in the United States in 2013 were 10% below 2005 levels, partly because of the replacement of coal by natural gas in the power sector since 2010.[17]
- While recognizing the benefits of natural gas over coal for carbon dioxide emissions, there have been questions on the net benefits on climate change because methane, another greenhouse gas (GHG), can escape during URD operations. However, investigators at the U.S. Department of Energy National Renewable Energy Laboratory recently found through a detailed meta-analysis that estimates of life cycle GHG emissions from shale gas-generated electricity are approximately half that of the average estimate of coal.[18]
- The use of water in energy production is an environmental sensitivity. Natural gas production requires less water than is required for producing the same amount of energy from coal, uranium, or biofuels. The amount of water needed to generate a million BTU is 0.6 to 1.8 gallons for natural gas, 1 to 8 gallons for coal, about 10 gallons for nuclear power, and 1,000 gallons on average for ethanol from corn.[19]

The overall benefits of URD appear to be numerous and obvious. Several federal government leaders have made positive statements about the benefits of URD for the United States, including Secretary of Energy Ernest Moniz (July 2014):

> The administration's view, my personal view, is that the gas revolution has had multiple benefits: CO_2 benefits, obviously economic benefits, jobs benefits. On the economic benefits, it includes lower prices.[20]

U.S. Environmental Protection Agency Administrator Gina McCarthy (December 2013) has been quoted as saying that natural gas has been a game-changer with our ability to move forward with pollution reductions that have been hard to get our arms around for many decades.[21] Secretary of the Interior Sally Jewel, a proponent of "fracking," believes that by using directional drilling and fracking, we have an opportunity to have a softer footprint on the land.[22]

ADDRESSING RISK

The rapid increase in URD has attracted significant public attention and increased scrutiny. This is evident by the huge increase in "fracking" articles in the print media since 2008, including anti- as well as pro-fracking documentaries and movies, ballot initiatives, and bans or moratoria enacted or discussed in localities where URD is possible (see Figure 11.2).

Figure 11.2
"Fracking" media coverage.

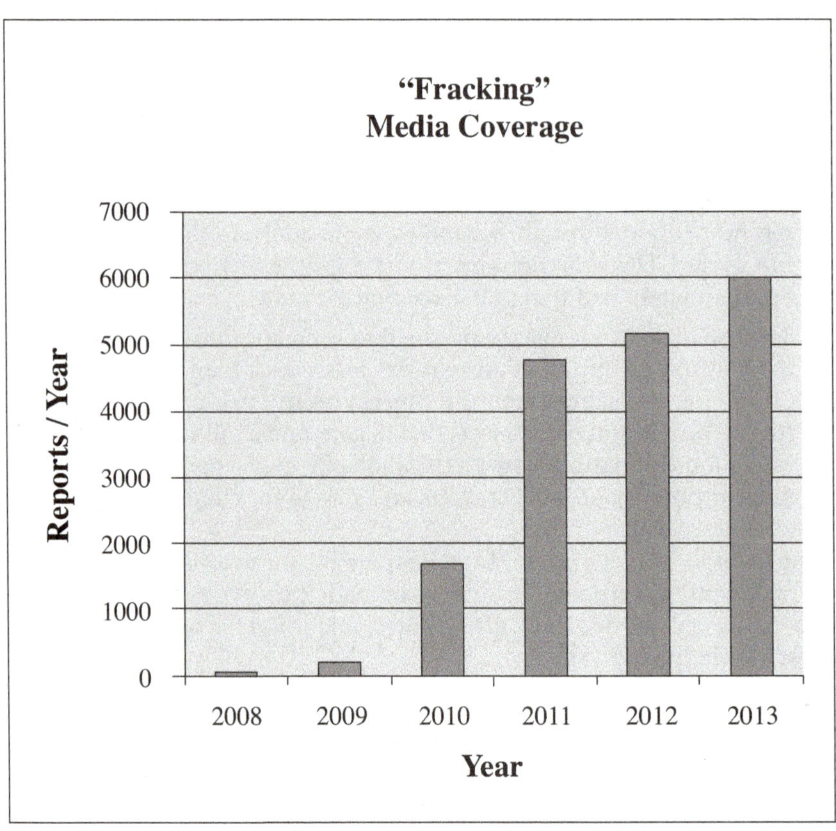

Source: Environmental Health News.[23]

The apparent reasons for opposition to URD are varied, ranging from citizens frustrated with industrial activity and appeals for a more precautionary approach to vehement opposition from those who view URD as a long-term threat to decarbonization of the global energy system.

As with any industrial activity, there is a trade-off between benefits and risks, and the industry generally strives to identify, understand, mitigate, and manage the risks of URD. This is done essentially by complying with regulations, adhering to long-standing risk management practices, and enhancing processes through innovation and research. URD operations are regulated at the local, state, and federal levels at every step of the process from site selection through reclamation, including review and approval of permits, well location and design, drilling operations, documenting and handling hazardous substances, water acquisition and disposal, air emissions, wildlife impacts, surface disturbance, worker health and safety, and inspections. There are at least nine major federal regulations that govern URD, with primacy generally delegated to state governments to execute the rules. In many cases, states have developed additional regulations. Furthermore, industry strives to go beyond regulatory requirements to manage risk by using and sharing proven engineering practices and seeking improvements (see Table 11.1; see also Sinding et al., Chapter 9 in this volume).

The American Petroleum Institute (API), through its American National Standards Institute accredited standards program, developed guidance documents to promote responsible practices by all operators in the field. In addition to more than 60 operating standards already in place for onshore exploration and production (including URD), API recently developed six new standards for URD relating to well construction and integrity, water management, practices for mitigating surface impacts, environmental protection, zonal isolation, and for community engagement.[24]

Many industry participants recognize that URD may occur near communities that are not familiar with industrial activity and that this activity may bring unsettling changes. Lights, noise, imposing structures, and increased traffic can temporarily affect the character of a community. Through early engagement with local officials and community members, industry representatives can establish a two-way conversation to discuss mutual goals for community growth, address concerns, and work to reduce potential impacts. A good example would be efforts to coordinate the timing and location of truck traffic to the extent possible. Industry can also discuss operational improvements that can mitigate the disruption to a community. For instance, how the use of multiwell pads and horizontal drilling can reduce the overall surface presence of URD operators by as much as 90%.

In a typical unconventional natural gas project, the most intensive activity occurs during drilling, which normally lasts between 4 to 6 weeks

Table 11.1
Examples of Responsible Development Strategies Used by URD Operators

Water Quality and Availability
- Determine predrilling area groundwater baseline water quality and evaluate potential water sources and disposal options
- Reuse flowback and increasing use of nonpotable water sources to reduce water use
- "Pitless" drilling; use of aboveground tanks for managing well fluids
- Closed loop drilling systems; all drilling fluid stored in steel tanks

Chemical Management
- Invest in fracturing fluids that use more environmentally benign components
- Carefully train employees to manage fluids according to established protocols

Air Quality and Reduced Emissions
- Use cleaner burning fuels or renewables to power on-site equipment, if practicable
- Emission mitigation technologies; e.g., green completion systems

Community Culture and Aesthetic
- Centralized water management systems, where practicable, that remove trucks from roads
- Sound control and surface management allow for safe drilling in close proximity to people
- Photovoltaic solar telemetry to transmit well data from remote locations to central office, reducing traffic and use of diesel fuels

Public Health
Collaborative industry initiatives to assess risks and support research

per well. The fracturing process itself typically lasts for only a few days for each well. Trucks drive to and from the site during the drilling and fracturing phase. Once drilling and fracturing are complete, all the temporary equipment is replaced with a wellhead, typically six to eight feet high, and there is little noise at the producing well. In addition to the wellhead, the site may also be used to locate small tanks and separators. The site remains in this state for the rest of its producing life.

Although the technologies used are highly effective and the goal is completely safe and responsible operations, rare accidental releases still occur. Through trade organizations such as the American Petroleum Institute and America's Natural Gas Alliance, members of the oil and natural gas industry strive to work together to eliminate all releases. Current technology to minimize risks includes wells with redundant layers of cemented steel piping to provide a shield between product and groundwater. Operators also use lined impoundments and storage tanks as barriers between wastewater, groundwater, and soil. Produced water, or flowback, is collected at the surface and recycled for future use or disposed of in a highly regulated process, typically through injection into deep disposal wells. Regarding emissions at the site, the New Source Performance Standards introduced by the EPA in 2012, to be implemented in early 2015, will lead to additional reductions by requiring "green completion" techniques that separate and capture gas and liquid hydrocarbons at the surface. The EPA estimates that these new rules will yield a nearly 95% reduction in volatile organic compounds (VOC) emissions from more than 11,000 new hydraulically fractured gas wells each year.[25] Furthermore, fugitive methane emissions should be substantially reduced through these green completions. On a broader scale that includes the gas distribution system, recent citywide studies that detected leaks from the gas pipeline infrastructure are prompting cities and gas suppliers to replace or repair aging pipes, which will also greatly reduce methane emissions over time.[26]

Stakeholders among industry, government, nongovernmental organizations, and the research community are engaged in multiple efforts to identify and implement new opportunities for reducing releases and eliminating impacts. Examples include new technologies to capture methane seepage and reduce fresh water use through desalinization of water or replacing it completely with gases or foams. Also, service companies are working to reduce or remove hazardous chemicals from the current fluids used in hydraulic fracturing.

Health Concerns

With the rapid expansion of URD, some public health advocates have raised concerns that not enough is being done to identify and assess potential health impacts, particularly regarding community exposures. Many members of industry meanwhile have recognized the need to focus additional attention on health concerns raised by the public. Both interests were addressed by the National Academies Institute of Medicine's Roundtable on Environmental Health Sciences, Research, and Medicine when it held a workshop on the Health Impact Assessment of Shale Gas Extraction in 2012.[27] The workshop brought together a range of experts to examine the state of the science, the direct and indirect environmental health impacts of shale gas extraction, and the prospects for further defining and minimizing potential health impacts.

The workshop concluded that the key health issues surrounding natural gas production from URD were respirable crystalline silica as an occupational hazard, undefined exposures via contamination of drinking water sources, community exposures to VOCs and ozone in ambient air, and psychosocial stress in communities near URD operations. Participants also identified challenges, including the difficulty of engaging polarized stakeholders to develop and share information, filling data gaps on the amount and type of emissions and discharges, and establishing baseline environmental concentrations and disease prevalence in local communities. They also questioned the prospective value of research when rapid advancements in technology and regulations are expected because these could appreciably alter relevant conditions and thus call the applicability of study findings into question.

Other parties and authoritative bodies have since undertaken substantial activities to better understand potential exposures and health impacts from URD, including numerous federal agencies (e.g., EPA studies on drinking water, methane, air emission standards, effluent guidelines, induced seismicity at disposal wells; Department of Energy chemical disclosure studies; Bureau of Land Management studies on disclosure, well integrity, water use, and flowback disposal; Occupational Safety and Health Administration studies on silica exposures and process safety; and the Department of Health and Human Services, which is funding a study to assess potential health impacts through the Centers for Disease Control and Prevention, National Institute of Environmental Health and Sciences, and National Institute of Occupational Safety and Health).

Professional scientific societies also are involved with assessing occupational and environmental health. Position papers and new committees opining on the state of science and research needs have been issued. Independent health-based institutes are engaged in planning collaborative research initiatives, including public–private partnerships, and researchers at many universities are securing research funds to investigate potential exposures and impacts.

Generally, these are qualified organizations that should be expected to conduct credible work on potential risks associated with URD—and credible work is clearly needed. Over the first few years of the substantial URD expansion, a range of reports on purported exposure pathways and health impacts have been produced with widely varying degrees of objectivity and scientific rigor. Furthermore, the media covers these reports with varying degrees of analysis and objectivity, often crossing the line between objective reporting and advocacy. Two respected science journalists have taken note of this, with one criticizing the practice of publicizing preliminary findings and the other commenting on the insistence of many in the news media to feed the hype machine by depicting every new study as "definitive and groundbreaking," diminishing scientific reporting to the point of being useless.[28]

The challenge for objective stakeholders is to critically assess all reports, regardless of funding source, to distinguish scientific, evidence-based findings from unsupported claims. This is particularly important for issues such as URD that are politically charged, with advocates willing to use exaggerated claims to draw support for their point of view. Although all claims of benefits and risks should be assessed for validity and accurately communicated to the public, this is especially true for those involving potential health impacts because an exaggerated or false statement regarding a health risk can lead to fear and chronic stress, which alone can have an impact on health.[29]

Along these lines, Ferrar et al.[30] analyzed the prevalence of self-reported health impacts and stressors in volunteers living near URD sites in the Marcellus Shale region. They found that stress was the most frequently cited of 59 unique health impacts, while "concern for health" was the predominant of 13 stressors. Other leading stressors were "concern that their complaints were being ignored" and that they were "being denied or provided with false information." The authors concluded that these stressors should be addressed as a priority by both government agencies and industry. However, this fear and stress could well be created, at least in part, by activists who promote unsubstantiated claims of significant harm regarding health effects as if they are firmly established facts. For example, one basic tenet of toxicology is that "the dose makes the poison"; therefore, all chemicals can be harmful at some dose, and most chemicals produce a pattern of toxic effects. A survey of any home, including the food pantry, would find a "long list of toxins" known to produce adverse effects, including those cited earlier in the chapter. Identifying inherent toxic hazards could be done for any list of chemicals, but without an understanding of potential exposures it does not define risk and can be quite misleading.

Could fear and stress have other specific consequences? A few reports cite adverse birth impacts associated with URD, but no related contamination has been reported.[31,32] If birth impacts are confirmed, could stress be a key factor? It is certainly prudent to inform the public about potential hazards, possible routes of exposure, and the probability of health risks associated with URD. Industry members and regulators have a lead role, presenting the information in the proper perspective. In contrast, promoting false or speculative health impacts as facts is irresponsible and unfair to those who cannot assess the information on their own. All URD stakeholders—governments, industry, activists, and the media—need to recognize this and work to mitigate stress as much as possible.

ASSESSING POTENTIAL HEALTH IMPACTS

What is needed, but thus far is mostly lacking, are studies conducted according to accepted scientific protocols. In particular, these should define the potential hazards originating from URD, a plausible route or routes by which humans are exposed, assessments of risk to health, and clinically confirmed adverse health effects that can plausibly result from those levels of exposure. Although chemicals present the most serious potential hazard, there can be other factors, such as those that induce psychological stress. Broadly speaking, risk assessors and risk managers in industry rely on four types of studies to help assess health concerns regarding URD.

1. Define Chemical Hazards

Hazard reports are helpful for identifying potential health effects. Examples include toxicity studies, or summaries of chemicals used in hydraulic fracturing fluids or in flowback waters. As stated earlier, reports that define chemical hazards have value in identifying potential effects, but they can be, and have been, misused to imply that the adverse effects identified are likely to be seen in communities near URD sites. Exposures are necessary for risk to exist.

2. Define Exposures

Exposure studies include monitoring of ambient air and air emissions from URD operations, as well as of drinking water resources in the vicinity of URD operations. To attribute the presence of chemicals in air or water to URD, it is important to first establish baseline conditions that exist before the URD activity. Thus far, air monitoring has been the more prevalent route of exposure studies, particularly over the Barnett Shale in Texas, which saw early development of URD. The Texas Commission on Environmental Quality (TCEQ) has led the way with a transparent, long-term monitoring program of ambient air quality, including the collection of 5 million data points since about 2000.[33] Interested parties can access the air quality data on the TCEQ website for criteria pollutants (e.g., particulate matter and ozone), hazardous air pollutants (e.g., benzene), and other VOCs. High-quality air monitoring projects in the Barnett Shale region include a study by the city of Fort Worth, which examined emissions associated with URD,[34] and a study by the Mickey Leland Center for Urban Air Toxics that examined contributions of emissions from active URD extraction and processing facilities to actual exposures in nearby residential communities.[35] However, even these reports are not without potential weaknesses; for example, the placement of monitors is not always optimal for measuring the impact of URD wells. Increased air monitoring is expected in the areas of all major U.S. shale plays due to new regulatory requirements, research plans (e.g., U.S. National Energy Technology Laboratory), and voluntary programs.

Public drinking water systems in the United States are currently monitored for more than 90 regulated contaminants, which should help to serve as a screening approach where URD is a concern. Private wells are not monitored to the same extent, but they have been monitored on a more ad hoc basis when URD created a concern, particularly in the Marcellus Shale region of Pennsylvania. Also, the U.S. Geological Survey has been conducting critical baseline groundwater monitoring in the Marcellus region finding naturally occurring methane in water from some wells in New York and Pennsylvania.[36,37]

Some groups are now focused on how to enhance environmental monitoring. One example is a collaboration of the National Resources Defense Council, the Harvard Center for Health and the Global Environment, the Mid-Atlantic Center for Children's Health and the Environment, and the Health Effects Institute that was launched with a workshop in December 2013.[38] This initiative is working to identify consistent, expert-vetted procedures for air and water monitoring that can be used to inform federal, local, and state stakeholders.

3. Risk Assessment

Assessing potential risk to health from URD should be done by integrating information on defined hazards with plausible routes of exposure. Exposure studies that use measured concentrations are most reliable regarding actual exposures, compared with modeled projections or the common current practice of reliance on individual's proximity to URD sites as a surrogate for exposure. One example of recent risk assessments is the study by Bunch et al.[39] They conducted an evaluation of potential health impacts from URD operations over the Barnett Shale region relying on more than 4.6 million air concentration values collected by the TCEQ. On the basis of a comparison of the air concentrations to federal and state health-based values, they concluded that the URD activities have not resulted in community-wide exposures that would pose a health concern. Clearly, more comprehensive studies to assess risk from actual URD-related exposures are needed to support risk managers.

4. Assessment of Clinical Evidence

The most reliable and comprehensive type of study for assessing actual health risks from a known source is one following standard epidemiological methods. Descriptive studies can assess trends in the incidence and prevalence of disease such as asthma, reproductive impacts, and cancer. The objective of these studies is to identify potential risk factors that could lead to disease. These studies are not definitive for a specific cause and effect. Analytical studies would need to be conducted to assess cause(s) and effect(s). Competent, robust epidemiological studies are usually expensive, time-consuming, and difficult to conduct, but they are needed to address concerns regarding URD. Furthermore, they must be conducted by competent investigators following established protocols. The characteristics most industry risk assessors look for in an epidemiological study are the following:

- Accepted study design (e.g., cohort, case–control, cross-sectional—the latter with acknowledged limitations regarding temporality of cause and effect)
- Properly selected exposed and unexposed groups (or cases and controls), with matching or stratification of potential confounders (e.g., age, socioeconomic status, smoking)
- Clinical documentation of health outcomes or another form of verification of health outcomes other than self-reporting
- Plausible exposure pathway scenario from source to receptor, verification of the plausibility of the pathway, and proper exposure metrics to describe the pathway
- Adequate control of potential selection bias, not self-selection
- Adequate control of potentially confounding variables
- Adequate statistical analysis, incorporating effects of confounding, interaction, temporality, coexposures, possible bias, and model selection, for example
- Adequate population sizes with proper documentation of precision (including confidence intervals)
- Adequate control of exposure classification bias, quantification of bias or sensitivity analyses
- Proper interpretation of results with strengths and weaknesses reflected, attention to internal consistency, coherence, multiple hypotheses testing, and alternative explanation of effects (e.g., coexposures, socioeconomic changes)
- Expert peer review and publication in a reputable journal

It is relatively rare that one study meets the full range and intent of these criteria, but they can be used to assess the general quality and credibility of epidemiological studies of potential health effects resulting from URD. That being said, community-based participatory research (CBPR) can be quite useful. CBPR studies are relatively inexpensive and quick to do. Although this type of research can have value in identifying issues to study that may not have been apparent, protocols must be followed to avoid several forms of potential bias or threats to study validity. One example of possible misuse is ascertaining exposure and outcome measures through self-reporting by individuals who are solicited or volunteer as participants because these may not be representative of the larger population. Lacking appropriate scientific protocols amounts to, essentially, anecdotal reporting. Although they may be helpful in identifying areas of concern and future study, they should not be viewed as definitive findings.

In summary, for any of the study types described earlier, the most reliable are those that are done using established protocols, are reviewed by peers with expertise in the field, and are published in reputable journals.

INDUSTRY ACTIVITIES REGARDING HEALTH CONCERNS

Community concerns and reports of adverse health effects associated with URD warrant attention and study. To address health-related issues, the American Petroleum Institute formed the Exploration and Production Health Issues Group. This multidisciplinary group of scientists is particularly focused on community health concerns for URD and uses an approach that emphasizes the following:

- Evidence-based scientific principles to identify, analyze, and respond to health-related concerns associated with all facets of URD
- Established methodologies to evaluate the risk posed by chemical components of fracturing fluids and flowback water, while emphasizing the importance of exposure information
- Development and support of public health research projects that provide valid findings for industry and government risk management decisions
- Collaboration with experienced scientists from industry, academia, and the government, possibly including public–private partnerships to identify, assess, and effectively manage potential health risks
- Transparent reporting of scientific findings, including peer-reviewed publications
- Clear, open communication to address public concerns and advance understanding of the benefits and risks of URD

Several projects sponsored by the American Petroleum Institute are under way or planned for the near future.

- Studies designed to quantitatively assess community exposure from URD operations and to evaluate whether a causal relationship exists between potential exposures and adverse health outcomes. These include three areas of focus: (1) identification and assessment of personal exposure to chemicals in people living near sites devoted to URD; (2) identification and assessment of sources of health effects data in the same community; and (3) completion of a formal epidemiological study utilizing the exposure and health effects data from the sources identified in phases 1 and 2 in the selected community. All study outcomes will be published in peer-reviewed journals.

- Using an approach similar to that just described, examine the existence, extent, source, and impact of psychosocial stress in communities near URD operations.
- Review the extent of possible exposures to naturally occurring radioactive material and any potential related health risk.
- Enhanced communication and outreach to stakeholders through public forums and publications.
- Address concerns for disclosure of hazards for chemicals in individual fracturing fluids using science-based processes.

These projects, and others to follow, will take considerable time to complete. In the meantime, industry members and their partners will continue with activities to identify and implement opportunities for reducing risk.

CHEMICAL DISCLOSURE AND HAZARD COMMUNICATION

The pumping of fracturing fluids, consisting of partially undisclosed composition, into well casings that pass through drinking water aquifers has raised concern. Several groups, including industry participants, are working to ensure that meaningful information regarding fracturing fluids is accessible to those who need it. However, this should be done while maintaining the confidentiality of proprietary chemicals. About 99% of fracturing fluid is water and sand, with 0.5% to 2% chemicals. Chemicals are used for multiple purposes, including preventing bacteria growth, reducing friction, and keeping the sand suspended in the solution. Specific chemicals imparting unique characteristics to fracturing fluids often provide a competitive advantage to the company involved. Companies have the right to protect the chemical identity from competitors as confidential business information (CBI), consistent with well-established U.S. law (e.g., Uniform Trade Secrets Act). Without this protection, there would be much less incentive for companies to spend funds to do research, innovate, and develop improved and potentially more sustainable fracturing fluids.

Having said that, the right to protect CBI does not release companies from the responsibility of communicating all hazards associated with their fracturing fluids. Industry should begin by disclosing as much of the composition as possible without compromising CBI. For some fluids, this means 100% disclosure. In cases in which precise identification of a chemical is not released, companies often provide the chemical category, which helps interested parties identify the general hazard. To help make compositional information broadly accessible, the Groundwater Protection Council, an organization of state regulators, and the Interstate Oil and Gas Compact Commission together created FracFocus.[40] FracFocus is a web-based registry where the nonconfidential chemicals used in the fracturing operation for an individual well site are posted for public access. In many states, the use of FracFocus is mandatory or allowed to meet requirements for disclosure of nonconfidential chemicals. Also, there are government-led initiatives on chemical disclosure, including work done by the Secretary of Energy Advisory Board Task Force on FracFocus 2.0 and the EPA.

Although a few constituents should remain confidential from public disclosure when appropriate to protect CBI, full chemical disclosure is made available to regulators, emergency responders, and physicians when circumstances warrant. Federal regulations under the Occupational Safety and Health Administration (OSHA) Hazard Communication Standard and the Emergency Planning and Community Right-to-Know Act (EPCRA) have explicit requirements that allow medical personnel access to CBI when needed to treat patients. In nonemergency situations, the medical provider may be asked to sign a confidentiality agreement (CA) before receiving the information, but in emergency situations the CA can be delayed. Also, OSHA and EPCRA require that Safety Data Sheets, which describe the health effects of hazardous chemicals in a product, be made available at the work site and to state and local officials and local fire departments when needed. In addition, states have developed or are developing public disclosure rules related to URD.

One notable example is a section of Pennsylvania's Act 13, which has been interpreted by some as a "gag order" on physicians.[41] However, a July 2014 ruling by a Pennsylvania Commonwealth court helped to clarify that it is not the intent of the law.[42] The ruling stated that nothing in Act 13 precludes a physician from including disclosed confidential and proprietary information in records given to another physician for the purposes of diagnosis or treatment or from including such information in a patient's medical record.

The proper balance between the legitimate need to keep certain constituents confidential and the public's right to know of any risks is one that can be met. Maintaining this balance is a common practice in many industry sectors where CBI exists. The risks associated with any fracturing fluid can be effectively communicated to the public without complete disclosure of the composition and can be achieved in most cases by reviewing the Safety Data Sheets that are mandated for fracturing fluids. Nonetheless, recognizing that a need may still exist, several industry-sponsored initiatives are considering tools to enhance descriptions of the overall hazard for the fluids.

CONCLUSIONS

Yes, there is a push to drill. In this environment, key questions that society must answer are the following: Is it wise to continue drilling, or should we wait? And, if we wait, for how long? Transparent, honest assessment of the benefits and potential risks can help to answer these (and other pertinent) questions. Reasonable people can disagree about the range of uncertainty around specific estimates, but it seems certain that unconventional natural gas and oil offer many substantial benefits of national, state, and local interest. These benefits have been described by numerous nonindustry experts in and out of government, are well documented, are currently being realized, and are expected to persist for decades to come.

Risks have also been identified, but in contrast to actual benefits, these are largely potential or hypothetical. Few examples of documented harm exist, and they are limited in scope. Still, industry participants know that certain risks exist, and there is a diligent effort in place to recognize, mitigate, and manage them. Furthermore, industry members recognize that some people are genuinely concerned about the impact of URD in their communities. Officials and risk communication experts report that this is not helped by a public discussion around "fracking" that is confused, biased, or ill informed. All sides must work to educate the public about the benefits and risks of unconventional oil and gas development. There is misinformation that must be clarified. Industry must continue to enhance its efforts to communicate effectively and transparently with all stakeholders. Over time, this will help address concerns and build trust.

ACKNOWLEDGMENT

I acknowledge and thank my colleagues at the American Petroleum Institute for their input and consultation—specifically, Patrick Beatty (API), Ziad Naufal (Chevron), Satinder Sarang (Shell), and Russell White (API).

NOTES

1. ExxonMobil Corporation. *Unconventional Resources Development* [video]. http://corporate.exxonmobil.com/en/community/corporate-citizenship-report/environmental-performance/unconventional-resources-timeline (accessed August 18, 2014).

2. ExxonMobil Corporation. *The Outlook for Energy: A View to 2040.* 2014. http://corporate.exxonmobil.com/en/energy/energy-outlook (accessed August 18, 2014).

3. International Energy Agency. *World Energy Outlook—2013, Part A: Global Energy Trends.* Paris: IEA Publications, 2013.

4. Lund S, Manyika J, Nyquist S, et al. *Game Changers: Five Opportunities for US Growth and Renewal.* McKinsey & Company, Insights & Publications. July 2013. http://www.mckinsey.com/insights/americas/us_game_changers (accessed August 18, 2014).

5. IHS Consulting. *America's New Energy Future: The Unconventional Oil and Gas Revolution and the U.S. Economy*. 2013. http://www.ihs.com/info/ecc/a/americas-new-energy-future.aspx (accessed August 18, 2014).

6. Ibid.

7. Bureau of Labor Statistics, U.S. Department of Labor. *Current Employment Statistics CES (National)*. http://.bls.gov/ces (accessed November 17, 2014).

8. Eisenberg R, Vice President Energy and Resources Policy, National Association of Manufacturers. Testimony of Ross Eisenberg on Opportunities and Challenges Associated with America's Natural Gas Resources. Senate Committee on Energy and Natural Resources. February 12, 2013. http://www.energy.senate.gov/public/index.cfm/files/serve?File_id=0ed60daf-4af5-4dda-a1ee-35c9a640aa7c (accessed August 18, 2014).

9. American Chemistry Council. U.S. chemical investment linked to shale gas reaches $100 Billion. February 20, 2014. http://www.americanchemistry.com/Media/PressReleasesTranscripts/ACC-news-releases/US-Chemical-Investment-Linked-to-Shale-Gas-Reaches-100-Billion.html (accessed August 18, 2014).

10. Kaskey J. U.S. shale spurs record investments by foreign chemicals. Bloomberg. June 26, 2014. http://www.bloomberg.com/news/2014-06-26/u-s-shale-spurs-record-investments-by-foreign-chemicals.html (accessed August 18, 2014).

11. Lund et al., *Game Changers*, op. cit.

12. U.S. EIA. *Short-term Energy Outlook*. August 12, 2014. http://www.eia.gov/forecasts/steo/index.cfm (accessed August 18, 2014).

13. Ibid.

14. Smith G. U.S. seen as biggest oil producer after overtaking Saudi Arabia. Bloomberg. July 4, 2014. http://www.bloomberg.com/news/2014-07-04/u-s-seen-as-biggest-oil-producer-after-overtaking-saudi.html (accessed August 18, 2014).

15. Stengel R, Scherer M, Jones R. Setting the stage for a second term. *Time*. December 19, 2012. http://poy.time.com/2012/12/19/setting-the-stage-for-a-second-term/4 (accessed August 18, 2014).

16. U.S. Energy Information Administration. Carbon dioxide emissions coefficients. February 14, 2013. http://www.eia.gov/environment/emissions/co2_vol_mass.cfm (accessed August 18, 2014).

17. U.S. Energy Information Administration. U.S. energy-related CO_2 emissions in 2013 expected to be 2% higher than in 2012. January 13, 2014. http://www.eia.gov/todayinenergy/detail.cfm?id=14571 (accessed August 18, 2014).

18. Heath GA, O'Donoughue P, Arent DJ, et al. Harmonization of initial estimates of shale gas life cycle greenhouse gas emissions for electric power generation. *Proceedings of the National Academy of Sciences of the United States of America* 11 (2014): E3167–E3176.

19. Mielke E, Diaz Anadon L, Narayanamurti V. *Water Consumption of Energy Resource Extraction, Processing and Conversion* (Energy Technology Innovation Policy Discussion Paper No. 2010–15). Belfer Center for Science and International Affairs, Harvard Kennedy School, Harvard University, October 2010.

20. Seifer B. EPA, Energy officials tout benefits of U.S. shale gas boom. *Energy in Depth*. July 31, 2014. http://energyindepth.org/national/epa-energy-officials-tout-benefits-shale-gas (accessed August 18, 2014).

21. Harder A. Why Obama should thank the oil and gas industry. *National Journal*. December 8, 2013. http://www.nationaljournal.com/power-play/why-obama-should-thank-the-oil-and-gas-industry-20131208 (accessed August 18, 2014).

22. Mills A. Denton's decision on fracking looms. *Tyer Morning Telegraph*, July 5, 2014. http://www.tylerpaper.com/TP-Business/202135/dentons-decision-on-fracking-looms#.U-p83Zoo7IU (accessed August 18, 2014).

23. Environmental Health News. Query: News stories, title contains fracking. http://www.environmentalhealthnews.org/archives.jsp?sm=fr4%3Btype6%3B5Story12%3BNews+Stories&tn=1title%2Clede%2Cdescription%2Csubject%2Cpublishername%2Ccoverage%2Creporter&tv=fracking&ss=1. (accessed August 18, 2014).

24. American Petroleum Institute. Standards. http://www.api.org/publications-standards-and-statistics/standards (accessed August 18, 2014).

25. U.S. Environmental Protection Agency. *Oil and Natural Gas Air Pollution Standards*. July 2, 2014. http://www.epa.gov/airquality/oilandgas/index.html (accessed August 18, 2014).

26. Zastrow M. Google maps methane leaks. *Nature News Blog*. July 17, 2014. http://blogs.nature.com/news/2014/07/google-maps-methane-leaks.html (accessed August 18, 2014).

27. Institute of Medicine. *Health Impact Assessment of Shale Gas Extraction: Workshop Summary*. Washington, DC: The National Academies Press, 2014.

28. Revkin A. Fracking bad for babies headlines are premature, study author says. Revkin.net. January 7, 2014. http://revkin.tumblr.com/post/72552907882/fracking-bad-for-babies-headlines-are-premature (accessed August 18, 2014).

29. Schneiderman N, Ironson G, Siegel SD. Stress and health: psychological, behavioral, and biological determinants. *Annual Review of Clinical Psychology* 1 (2005): 607–628.

30. Ferrar KJ, Kriesky J, Christen CL, et al. Assessment and longitudinal analysis of health impacts and stressors perceived to result from unconventional shale gas development in the Marcellus Shale region. *International Journal of Occupational and Environmental Health* 19: (2013): 104–112.

31. McKenzie LM, Guo R, Witter RZ, et al. Birth outcomes and maternal residential proximity to natural gas development in rural Colorado. *Environmental Health Perspectives* 122 (2014): 412–417.

32. Whitehouse M. Study shows fracking is bad for babies. *Bloomberg View*. January 4, 2014. http://www.bloombergview.com/articles/2014-01-04/study-shows-fracking-is-bad-for-babies (accessed August 18, 2014).

33. Texas Commission on Environmental Quality. http://www.tceq.state.tx.us (accessed November 18, 2014).

34. Eastern Research Group Inc. and Sage Environmental Consulting LP. *City of Fort Worth Natural Gas Air Quality Study*. July 13, 2011. http://fortworthtexas.gov/uploadedFiles/Gas_Wells/AirQualityStudy_final.pdf (accessed August 18, 2014).

35. Zielinska B, Fujita E, Campbell D. *Monitoring of Emissions from Barnett Shale Natural Gas Production Facilities for Population Exposure Assessment* (Final Report to the Mickey Leland National Urban Air Toxics Research Center, No. 19). 2011. https://sph.uth.edu/mleland/Webpages/publications.htm (accessed August 18, 2014).

36. Kappel WM, Nystrom EA. *Dissolved Methane in New York Groundwater, 1999–2011* (U.S. Geological Survey Open-File Report 2012-1162). August 2012. http://pubs.usgs.gov/of/2012/1162 (accessed August 18, 2014).

37. Senior LA. *A Reconnaissance Spatial and Temporal Assessment of Methane and Inorganic Constituents in Groundwater in Bedrock Aquifers, Pike County, Pennsylvania, 2012–13* (U.S. Geological Survey Scientific Investigations Report: 2014-5117). 2014. http://pubs.er.usgs.gov/publication/sir20145117 (accessed August 18, 2014).

38. Center for Health and the Global Environment at Harvard, Mid-Atlantic Center for Children's Health and the Environment, Health Effects Institute, and Natural Resources Defense Council. NRDC Workshop to Develop Recommendations for Environmental Monitoring Related to Unconventional Oil and Gas Extraction, December 12–13, 2013; Workshop Proceedings; June 2014. http://www.nrdc.org/health/14053001.asp (accessed August 18, 2014).

39. Bunch AG, Perry CS, Abraham L, et al. Evaluation of impact of shale gas operations in the Barnett Shale region on volatile organic compounds in air and potential human health risks. *Science of the Total Environment* 468–469 (2014): 832–842.

40. FracFocus2.0. http://fracfocus.org (accessed August 18, 2014).

41. Jewell, FC. Oil and gas industry needs to better inform public debate on fracking. *National Journal*. November 8, 2013. http://www.nationaljournal.com/energy/jewell-oil-and-gas-industry-needs-to-better-inform-public-debate-on-fracking-20131108 (accessed August 18, 2014).

42. State impact.npr.org/Pennsylvania/2014 (accessed August 18, 2014).

Chapter 12

The Geology and Sustainability of Shale

J. David Hughes

INTRODUCTION

After decades of declining oil and gas production in the United States, the so-called shale revolution has allowed production to grow once again. The revolution was made possible by technological innovations allowing hydraulic fracturing to be combined with horizontal drilling, which unlocked oil and gas resources in low permeability reservoirs that had previously been inaccessible. It began in eastern Texas with the development of natural gas production from the Barnett Shale in the late 1990s and quickly spread to shale gas plays in Louisiana, Oklahoma, Arkansas, Pennsylvania, and other states. The technology was then applied to oil with the development of the Bakken Shale in North Dakota and Montana and later the Eagle Ford Shale in southern Texas, as well as other plays. The shale revolution has generated a great deal of enthusiasm on the future of oil and gas production in North America with talk of U.S. "energy independence" and "Saudi America."[1] Globally there is hope that shale will provide a new energy bounty. Given such zeal, it is useful to examine some of the fundamental characteristics of shale gas and oil production, along with some of the assumptions, with a view to determining its long-term sustainability.

CONTEXT

Notwithstanding the concerns about climate change, fossil fuels made up 87% of U.S. and global energy consumption in 2013.[2] Of this, oil and gas accounted for 66% of U.S. energy consumption and 57% of global consumption, with the balance being made up by coal. Nonhydropower renewables, such as solar, wind, biomass, and geothermal, made up just 2.6% of U.S. energy consumption and 2.2% globally in 2013. Forecasts suggest that world energy consumption will grow by 50% through 2040, at which time oil and gas will still make up 56%.[3] So, if one believes the forecasts, oil and gas will be important contributors to the global energy mix out to 2040 and likely considerably beyond.

THE NATURE OF SHALE RESERVOIRS

Oil and gas are generated from organic matter in sediments buried to depths at which temperatures are sufficient to convert the organic matter, over millions of years, first to oil, and then, with even higher temperatures, to gas. Such organic rich sediments are termed "source rocks." In conventional production, oil and gas migrates from the source rock to a reservoir where it is trapped by an impermeable seal, whether structural, through folding or faulting, or stratigraphic, as a result of lateral changes in rock composition. Conventional reservoirs typically have sufficient porosity and permeability so that oil and gas can migrate considerable distances to vertical well bores where they are produced.

Shale reservoirs, in contrast, are the source rocks themselves, which contain oil and gas that has not migrated significantly due to the extremely low permeability of the rock. Although shale is the generic term, some of these reservoirs may be carbonates or other lithologies with the common characteristic being that they are very "tight," meaning they have extremely low permeability. Hence the term "tight oil" as opposed to "shale oil."

The only way to extract oil and gas from shale reservoirs is by inducing artificial permeability through the process of hydraulic fracturing. Although the oil and gas industry has been stimulating wells to enhance permeability for many decades, the scale at which it is now practiced in conjunction with horizontal drilling is new and still evolving. Horizontal laterals can follow a thin shale bed for considerable distances and contact the reservoir for several thousand feet, as opposed to vertical wells, which may only contact the reservoir for a few tens of feet. Coupled with hydraulic fracturing, this makes all the difference between an uneconomic well and a well that can produce commercial volumes of oil and gas.

Hydraulic fracturing ("fracking") involves the injection of large volumes of water under extreme pressure to fracture the reservoir, along with a proppant (typically sand but ceramic beads or other materials can also be used) to hold the fractures open, and other chemicals to facilitate the process. Typically only a small portion of the horizontal lateral is treated at a time, which is termed a "stage." Depending on its length, a horizontal lateral may have 30 or more stages. Typical wells use on the order of 5 million gallons of water; however, some of the largest frack jobs in northeastern British Columbia averaged more than 16 million gallons per well.[4]

All shale reservoirs are not created equal, and even within productive shale reservoirs there can be a considerable variation in quality. Important parameters include organic matter content, thermal maturity, composition (which can enhance brittleness to propagate fractures), thickness, and the presence of natural fractures. The "best plays," such as the Bakken and Eagle Ford for tight oil and the Marcellus for shale gas, cover large areas, typically several thousand square miles, although the "sweet spots" or "core areas" tend to be confined to a small portion of the total play areas.[5]

DISTRIBUTION AND CURRENT PRODUCTION

Although the distribution of shale plays in the lower 48 U.S. states appears to be widespread, as of mid-2014, 48% of shale gas production came from just two plays—the Marcellus and Eagle Ford plays—and 78% came from the top five plays. In the case of tight oil, 62% of mid-2014 production came from two plays—the Bakken and Eagle Ford—and 84% came from the top five plays—Marcellus Shale, Haynesville Shale, Barnett Shale, Utica Shale, and Woodford Shale.[6]

Figure 12.1 illustrates current shale gas production in the U.S. by play. Production of just over 36 billion cubic feet per day (bcf/d) in July 2014 amounted to slightly more than half of total U.S. dry gas production.[7] Production has essentially grown substantially since 2006. The long-term sustainability of production from these plays, however, is in question. For example, the Barnett play, which was the first shale gas play to employ high-volume, multistage, hydraulic fracturing of horizontal wells, peaked in 2011 and is now down 18% from peak. The Haynesville play, which was the largest shale gas play in the United States when it peaked in early 2012, is now down 46%. Other plays like the Fayetteville and Woodford, which were unknown in 2006, are on a gently declining plateau. Five legacy plays, which produced 58% of U.S. shale gas production at their collective peak in August 2012, are now down 23%. Growth in shale gas production is now primarily being supported by the Marcellus and associated gas from the Eagle Ford, with less significant production from the Bakken, Utica, and a handful of other small plays, which will also reach their peaks over the next few years.

With regard to current tight oil ("shale oil") production in the United States, two plays stand out, the Bakken and Eagle Ford, which grew from

Figure 12.1
U.S. shale gas production by play from 2000 through 2014.

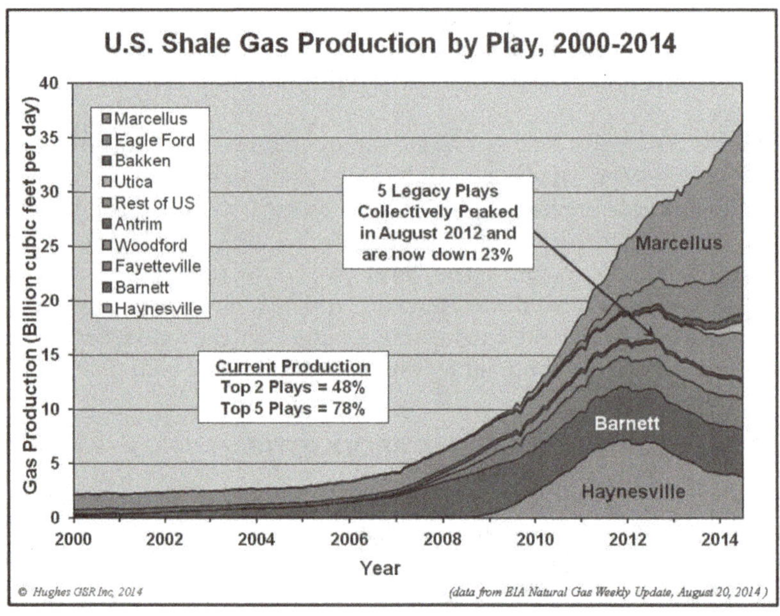

essentially nothing in 2006 to production of more than 2 million barrels per day (mbd), which is more than a quarter of total U.S. crude oil production. The other plays, several of which lie in the Permian Basin of Texas and New Mexico, as well as in other central states such as Colorado, Oklahoma, and Wyoming, are much less spectacular. Most of these are not new finds at all but rather represent redevelopment of old plays with new technology; most produced significant amounts of oil in 2000, well before the onset of the "shale revolution." There are also minor contributions of liquids production from shale gas plays such as the Barnett and Marcellus. As with shale gas, the long-term sustainability of oil production from these plays is in question.

FORECASTS

The Energy Information Administration (EIA), the main source of energy statistics and forecasts within the U.S. government, is bullish on the future growth of shale gas and tight oil production in the United States. The EIA's latest shale gas projections in several cases are illustrated in Figure 12.2. In its reference case, the EIA expects shale gas production to more than double from 2013 levels and comprise more than half of total U.S.

Figure 12.2
Energy Information Administration scenarios of U.S. shale gas production through 2040.

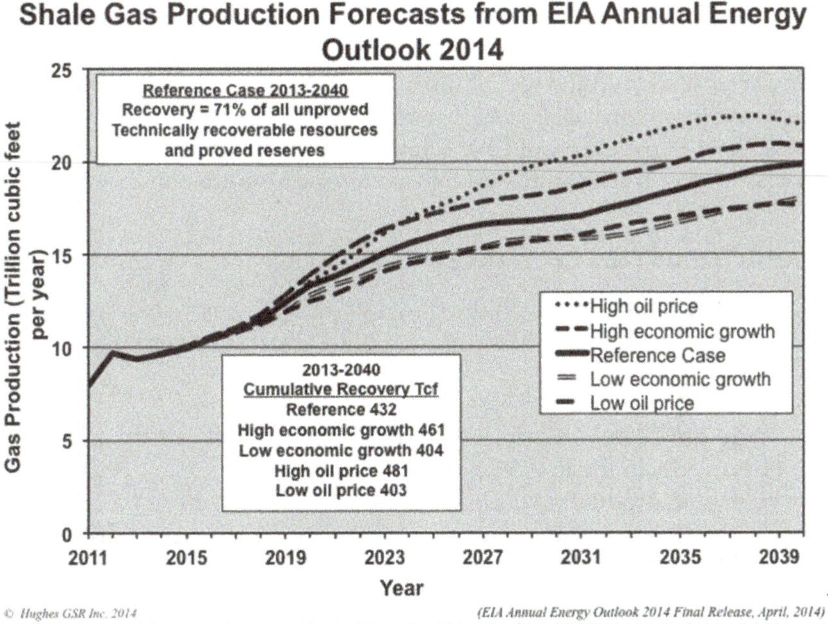

Source: Energy Information Administration. *Natural Gas Weekly Update.*[8]

gas production in 2040. These forecasts assume the consumption of between 66% and 79% of the EIA's estimated 611 trillion cubic feet of unproved shale gas resources and proved reserves by 2040. Unproved resources, which are the bulk of what is thought to be technically recoverable, have no implied price required for extraction and are highly uncertain compared with proved reserves, which are recoverable with current technology under current economic conditions. Unproved technically recoverable resources are estimated by the EIA at 489 trillion cubic feet and proved reserves at 122 trillion cubic feet, so these scenarios amount to the recovery of 66% to 79% of all proved reserves and unproved resources by 2040.

The EIA is similarly bullish on future production of oil from shale. EIA's projections for tight oil production assume the extraction of between 37 billion (low oil price case) and 47 billion barrels (high oil price case) by 2040. This amounts to all of the 7.15 billion barrels of existing proved tight oil reserves and between 50% and 67% of the EIA's estimated 59.2 billion barrels of unproved tight oil resources.[9]

Because the EIA forecasts conventional oil and gas production to maintain a plateau or decline, shale is projected to be the only source of significant growth, estimated to become more than half of total production by 2040.[10] To meet its forecasts, the EIA projects the drilling of more than 1.5 million oil and gas wells between 2014 and 2040 at rates exceeding 60,000 wells per year toward the end of this period. The questions become, how realistic are these forecasts, and what are the implications of being wrong? An ancillary question would be: what are the collateral environmental consequences of trying to maintain these forecast production levels?

CHARACTERISTICS OF SHALE PRODUCTION

A detailed analysis of well production data for the major shale plays reveals several characteristics in common with which to determine future production levels.[11]

1. There are high well production declines, typically in the range of 80% to 85% in the first 3 years.
2. High field production declines occur as well, typically in the range of 23% to 49% per year, which must be replaced with more drilling to maintain production levels. This compares to static field declines in the range of 5% to 6% per year in major conventional oil fields and 20% to 25% for conventional gas fields.
3. Plays are not uniform and invariably have "core" areas or "sweet spots" (i.e., where individual well production is highest and hence the economics are best). Sweet spots are targeted and drilled off early in a play's life cycle, leaving lesser quality rock requiring higher oil and gas prices to be drilled as the play matures. Thus, the number of wells required to offset field decline progressively increases with time.
4. The plays are finite in area and therefore have a limited number of locations to be drilled. Once the locations run out, production goes into terminal decline.
5. The rate of production is directly correlated with the rate of drilling, which requires constant high levels of capital investment, given that wells cost $3 million to $9 million each.
6. The widespread public pushback at the environmental issues surrounding fracking has limited access in states like New York and Maryland and several municipalities and has also triggered lawsuits. This is likely to continue and further limit access to the tens of thousands of drilling locations required to maintain, let alone grow, production.

A review of major shale plays reveals that although some are still increasing in production, such as the Bakken, Eagle Ford, and Permian Basin for tight oil and the Marcellus for shale gas, others, such as the Barnett, Haynesville, Fayetteville, and Woodford, are in decline.[12] Taking a long-term approach, it is almost certain that all shale plays go through the same cycle; plays such as the Haynesville have gone from discovery to the largest shale gas play in the United States to steep decline in only 7 years. It is not impossible to reverse declines in plays like the Haynesville with higher levels of investment and drilling, it is just unlikely that that will happen unless prices go considerably higher. Similarly, in the other plays, prices will have to go considerably higher to justify drilling in lower quality rock when locations in sweet spots run out. The EIA meanwhile is suggesting relatively low prices in its reference case for the foreseeable future (less than $6/MMBtu gas and $115/barrel oil out to 2030).[13] These prices are highly unlikely to support its production forecasts from shale in the longer term.

Projecting future production in shale plays is a matter of determining essential play parameters such as field decline rate, well quality by area, and number of available drilling locations. Then, by estimating drilling rates, a future production profile can be developed.

A brief case study[14] of the oldest shale gas play, the Barnett, serves to illustrate the salient fundamentals that ultimately control long-term production from all shale plays:

- Play area was determined by examining production data from all wells targeting the Barnett Shale reservoir. The play limits were delineated by wells with limited or no production. Total play area was estimated at 5,140 square miles.
- Field decline rates were calculated from existing drilling data on a county-by-county basis using a commercial database of well production data by vintage. These range from 19% to 25% with a mean of 23%.
- Well quality was determined using average production rates over the first 12 months of well life on a county-by-county basis. These range from 308,000 cubic feet per day (mcf/d) in outlying counties to 1,740 mcf/d in the core "sweet spot." From the field decline and well quality, the number of new wells required per year required to keep production flat is determined to be 1,160.
- The number of remaining drilling locations was determined by assuming final well density will be eight wells per square mile, so coupled with wells already drilled, 21,788 locations remain, for a final total of 41,462 wells when the play is fully developed (including some 3,732 wells drilled that are no longer producing).

Figure 12.3
Future production of the Barnett shale gas play assuming various drilling rate scenarios.

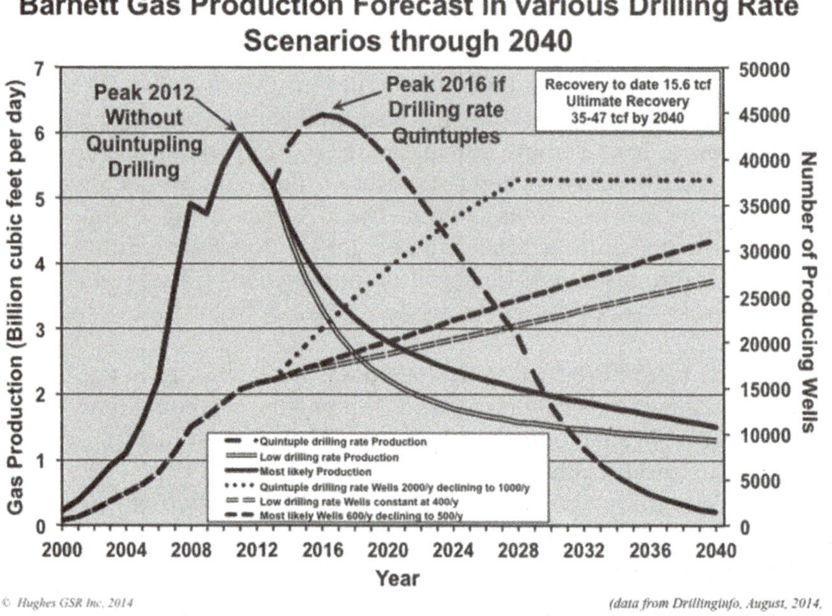

Source: Energy Information Administration. *Annual Energy Outlook 2014.*[15]

Drilling rates covered a range of scenarios, ranging from the continuation of current drilling rates (low drilling rate scenario), to significantly ramping up drilling rates assuming prices will go higher to justify drilling in lower quality parts of the play. These scenarios are illustrated in Figure 12.3.

The veracity of these production projections for the Barnett can be validated by comparing the gas recovery through 2040 to that projected by a detailed study conducted by the Bureau of Economic Geology and the University of Texas at Austin (UT study).[16] The future production scenarios in Figure 12.3 suggest that between 35 and 47 tcf will be recovered by 2040 depending on drilling rate, whereas the UT study projects a 45 tcf ultimate recovery (assuming considerable gas will be recovered beyond 2040).[17]

Several things are clear from this analysis:

- Significantly increasing production will require considerably higher prices to justify the drilling rates required.

- Drilling at higher rates does not significantly increase ultimate recovery, although recovering the gas sooner will make the supply picture worse later.
- More than double the current number of wells will need to be drilled in the play to meet these projections.

Similar analyses of the other major shale gas plays reveal a comparable pattern, although on different time scales. They point out that gas production in the longer term will be difficult to maintain from known plays and will require higher prices to justify the drilling of ever larger numbers of lower-quality wells to prop up production. Emerging shale gas plays will be required to meet huge growth rates to meet EIA expectations for its production forecasts, which is unlikely. In the case of tight oil, production projections of the "most likely" drilling rate scenarios for the two major plays, the Bakken and Eagle Ford, suggest production will continue to grow to a peak in 2016 at a combined production level of 3.1 mbd followed by a decline to about 0.1 mbd by 2040. As with shale gas, emerging plays will be required to meet huge growth rates to meet the EIA's forecasts. A complete review of major shale plays with production projections is beyond the scope of this chapter but is presented in a separate report.[18]

IMPLICATIONS

Although shale gas and tight oil have been temporary game-changers in that they have reversed declines that were thought to be terminal as recently as 2005, long-term sustainability is highly questionable. Assumptions about future domestic supply of oil and gas are critical components of energy policy, and current forecasts are highly likely to be overstated.

The belief in cheap and abundant gas over the long term has created an inelastic growth in demand for electricity generation and petrochemicals. Investment has been diverted from alternative sources of energy, raising vulnerability to cost hikes and supply interruptions. Moreover, there is pressure to expedite liquid natural gas (LNG) exports of U.S. gas on the assumption of long-term growing supplies. In my opinion, this can only put pressure on domestic prices and future energy security.

Similarly, the perceived long-term abundance of tight oil is putting pressure on U.S. politicians to relax the decade's old ban on exporting crude oil.[19] This is certainly understandable from a short-term corporate perspective but is ill advised for a comprehensive energy policy looking at long-term energy security.

Meeting the EIA forecasts for oil and gas production through 2040, of which more than half will come from shale, assumes the drilling of more than 1.5 million new wells. Although there is no free lunch with any energy source—all have their environmental footprint—shale development has been a source of particular controversy. Trying to meet expectations for shale production will mean the environmental impacts experienced to date are only the beginning. Many more wells will have to be drilled in known plays over the next two decades concomitant with intense drilling in emerging plays. U.S. energy policy would be well advised to factor the realities of shale into its long-term energy planning. Shale certainly has been a short-term game-changer, but long-term realities must be considered in a sustainable energy plan.

NOTES

1. "Saudi America." *The Economist* [print edition], February 15, 2014. http://www.economist.com/news/united-states/21596553-benefits-shale-oil-are-bigger-many-americans-realise-policy-has-yet-catch (accessed August 20, 2014).

2. *BP Statistical Review of World Energy 2014*. http://www.bp.com/en/global/corporate/about-bp/energy-economics/statistical-review-of-world-energy.html (accessed August 20, 2014).

3. Energy Information Administration. *International Energy Outlook 2013, Reference Case*/ http://www.eia.gov/forecasts/ieo (accessed August 20, 2014).

4. B.C. Oil and Gas Commission. *Investigation of Observed Seismicity in the Horn River Basin*. August 2012. http://www.bcogc.ca/node/8046/download (accessed August 20, 2014).

5. Energy Information Administration. *Natural Gas Weekly Update*. http://www.eia.gov/naturalgas/weekly (accessed August 23, 2014).

6. Ibid.

7. Energy Information Administration. *Annual Energy Outlook 2014. Assumptions*. http://www.eia.gov/forecasts/aeo/assumptions/pdf/oilgas.pdf (accessed August 23, 2014).

8. Ibid.

9. Energy Information Administration. *Annual Energy Outlook 2014*. http://www.eia.gov/forecasts/aeo (accessed August 23, 2014).

10. Ibid.

11. Hughes JD. 2013. *Drill, Baby, Drill: Can Unconventional Fuels Usher in a New Era of Energy Abundance?* Post Carbon Institute. http://assets-production-webvangta-com.s3-us-west-2.amazonaws.com/000000/03/97/original/reports/DBD-report-FINAL-pdf (accessed August 20, 2014).

12. International Energy Agency. *World Energy Outlook 2008*. http://www.worldenergyoutlook.org/media/weowebsite/2008-1994/weo2008.pdf (accessed August 20, 2014).

13. Ibid.

14. Hughes JD. *Drilling Deeper on Shale*. Post Carbon Institute. October 2014. http://www.postcarbon.org/wp-content/uploads/2014/10/Drilling-Deeper_FULL.pdf (accessed November 27, 2014).

15. Energy Information Administration. *Annual Energy Outlook 2014*. http://www.wia.gov/forecasts/aeo (accessed August 20, 2014).

16. Hughes, *Drilling Deeper on Shale,* op. cit.

17. Browning J, Tinker SW, Ikonnikova S, et al. Barnett study determines full-field reserves, production forecast. *Oil and Gas Journal.* http://www.ogj.com/articles/print/volume-111/issue-9/drilling-production/barnett-study-determines-full-field-reserves.html (accessed August 23, 2014).

18. Hughes, *Drilling Deeper on Shale,* op. cit.

19. IHS. *Crude Oil Export Special Report. U.S. Crude Oil Export Decision. Assessing the Impact of the Export Ban and Free Trade on the US Economy.* http://www.ihs.com/info/0514/crude-oil.aspx (accessed August 23, 2014).

Chapter 13

Epilogue

Anthony R. Ingraffea

An epilogue to a book is supposed to serve as a comment on or a conclusion to a written work. It should summarize that which has been written and offer thoughts for the future. In short, an epilogue should be presented as a perspective and wrap things up. My summary of the chapters you have read is quite simple: even now, a decade or so into the shale revolution, we know far less than we should about its various environmental, health, and climate effects. What we do know indicates that all of those effects are neither negligible nor trivial. Despite the warnings from scientists, health professionals, and engineers who call for more research to assess the safety of the process, hydraulic fracturing in shale has proceeded unabated. To date, only a few local efforts have been successful in stopping or slowing down the development of unconventional gas extraction (UGE). For many reasons this is depressing, because the stakes are so high. I explain what I mean by this in the following paragraphs.

Five years ago, few empirical data were available to help answer questions about this new technology. Word of mouth sometimes brought answers to questions from people around the world who wanted to know about the impacts of unconventional gas extraction. Now we have entire journals, books like this one, and extensive online resources to help answer questions. I hope that this book has helped to inform the reader of the consequences, pro and con, of UGE, and to illustrate that UGE is equivalent to the oil and gas industry blundering into the deepest, darkest corners of its global hydrocarbon warehouse. Given the marvels of technology, the oil and gas industry jumped on the natural gas extraction bandwagon to bring to market the resources locked deep in shale formations. Shale had been essentially off limits to the industry because of its extremely low permeability. Basically it was not cost-effective to extract natural gas or, more recently, oil from shale. Until now.

Shale rock is thousands of times less permeable than traditional source rocks for hydrocarbons. Essentially, the flow of hydrocarbons through the rock itself is too slow for shale to be a useful source rock. In contrast, a shale rock mass can be made acceptably permeable through a form of well stimulation, hydraulic fracturing (colloquially termed "fracking"), which reopens existing joints (think of these as nearly vertical natural cracks) and cleats (think of these as nearly horizontal natural layer separations) in the shale rock mass. The process allows for the extraction of oil and gas thousands of feet below the surface. Many of the authors in this book have discussed the extraction of natural gas from shale and have commented that conventional drilling has been going on for decades. It is unconventional gas extraction—fracking in shale—that has prompted the huge debate about the safety, economic value, and wisdom of embracing this technology without having data on which to base decision making. We know a little bit about the insult to the environment from UGE and far too little about the health effects both short- and long-term.

UGE creates marginally acceptable productivity only when accompanied by *technologies of scale* and by application of *spatial intensity* in an oil/gas play. Low permeability drives up well scale and spatial intensity, and these in turn drive increased risk. Technologies of scale include, for example, the use of 50 to 100 times more fracturing fluid in a shale well than in a nonshale well. The average Marcellus Shale gas well consumes about 5 million gallons of frack fluid; some shale wells in other plays consume five times as much. More fracturing fluid in turn requires more water, more sand, and more chemical additives and produces more fluid waste. The process from drilling the well, extracting the gas, transporting the gas, and storing the frack fluid creates environmental disruption, imposes a high demand on road/bridge infrastructure, and creates a higher risk for transport and pad spills of dangerous substances and toxic waste disposal problems. Technologies of scale also include the use of much longer wells, with lateral length often exceeding well vertical depth. Longer wells require heavier drilling equipment and longer drilling periods and cause more challenges for successful cement jobs. These create higher risk for unacceptable local air, noise, and light pollution as well as contamination of underground sources of drinking water.

Oil or gas is present in variable concentrations throughout a shale play—that is, it is not localized into "pools." Therefore, to maximize extraction of the oil or gas, the process involves drilling many wells in the play. Ideally, drilling pads are distributed in a checkerboard-like pattern on the surface, and each pad services a production or spacing unit, usually a rectangle-like surface area of about a square mile. Each pad has many wells, typically eight or more. Directional drilling then allows half of these wells to have laterals running in one direction, the other half in the opposite direction. The laterals are run roughly parallel and are separated by a few hundred feet. In the Marcellus shale play, for example, each well would "drain" about 80 surface acres; well density is about 8 wells per square mile. This is the meaning of "spatial intensity."

The principal consequence of spatial intensity is that homeowners, farms, schools, and businesses are required to coexist within a widespread, heavy industrial zone that includes a complex of drilling pads, storage tanks, compressor stations, processing units, and pipelines. Setback distances from pads and from ancillary infrastructure and private water wells become important to minimize harm to individuals and animals.

The necessity for wide-scale drilling to extract an economically acceptable amount of gas from low rock permeability is essentially a heavy-handed, widespread bludgeoning of a shale play for an acceptable return of hydrocarbons. There is a tremendous amount of disruption to the environment, below and above ground, to extract a sufficient amount of hydrocarbons. Many have questioned the value of the effort, while others have embraced it as the "best" way to "solve" the nation's energy needs. Yet both sides present cogent arguments absent data from well-designed studies. Chapters in this book present the most recent evidence on the pros and cons of UGE; however, each of the authors call for more well-designed studies to provide answers to the questions of potential harm to the environment and to animal and human health.

The evidence on UGE's effect on climate change, primarily due to the "methane effect," for example, is at a more advanced stage and by all indications should give us pause for concern. There are serious issues that must be addressed in an empirical, nonpolitical, nonjudgmental way before we, here and abroad, embrace UGE as a "clean, efficient" alternative to coal and oil. To do otherwise would be highly irresponsible and potentially irreversibly damaging to the environment and to human and animal health. The million-dollar questions include the following: Can this natural resource be extracted in a safe and economical way? Is UGE worth the effort? Is there a win–win solution?

Proponents of UGE, including those in the oil and gas industry, government, and special interest groups all say that when it comes to extracting natural gas from shale, they want to "get it right." What does it mean when industry and government entities claim that they want to get it right? I interpret "getting it right" to mean, regardless of cost to the industry, there must be adherence to three fundamental principles. All regulatory quality judgments should derive from these three fundamental principles behind truly tough regulations:

1. The immediate and cumulative negative impacts on environment, human and animal health, and climate change from shale gas/oil development must be assessed and acknowledged.
2. Taxpayers should not be responsible for costs and damages created by the shale gas/oil industry.
3. There must be a net socioeconomic gain from shale gas/oil development.

With these principles consistently in mind, there is no need to dive deeply into any particular regulatory element. One need only determine whether a proposed regulation addresses one or more of these basic principles and then determine whether the regulatory agency "got it right" without regard to cost to the gas/oil operator. For example, is it really meaningful to argue whether cementing of surface casing should stop 30 feet or 120 feet below the suspected depth of fresh water when it is well known that no length of cementing will guarantee a well will not leak into an underground source of drinking water? Yes, it might be meaningful, but only if you are trying to decrease the probability of such contamination, knowing that that probability can never be zero. But a truly tough regulation on this issue should simply require:

1. Cementing to surface of all conductor, surface, and intermediate casing strings.
2. Monitoring of wells for leakage outside the production casing indefinitely.
3. Immediate repair, if possible, of any leaking well.
4. Immediate restoration of an equally convenient freshwater supply to affected parties during repair and possibly indefinitely if repair is unsuccessful.

When one makes comparisons between existing so-called tough regulations and truly tough regulations, one sees that the existing regulations have been negotiated to relieve the operator of increased cost and risk of litigation while increasing the cost and risk to the public.

This general guideline for truly tough regulation of shale oil/gas development should make it clear that so-called tough regulations are almost always unbalanced, wordy, opaque compromises between industry operator expenses and a socioeconomic and environmental negative impact. There has been no clearer statement of this fact than the wording in a recent Colorado court decision regarding the ability of the City of Longmont to prohibit shale gas/oil development from its residential areas:

> While the Court appreciates the Longmont citizens' sincerely-held beliefs about risks to their health and safety, the Court does not find this is sufficient to completely devalue the State's interest, thereby making the matter one of purely local interest. . . . Longmont's ban on hydraulic fracturing does not prevent waste; instead, it causes waste. Because of the ban, mineral deposits were left in the ground that otherwise could have been extracted. . . . Mineral deposits are being left in the ground by all the wells that are not being drilled due to the fracking ban.[1]

Existing regulations, more often than not, are pathways to a low-threshold legal responsibility and low financial risk for the industry negotiated by the industry, legislators, regulators, and, sometimes, environmental organizations. This negotiation is usually flavored with warnings about loss of opportunity for tax revenue and jobs, while the scientific bases for counterarguments about health, economics, and climate are incomplete or entirely unavailable, as you probably discovered by reading the chapters in this book.

I live and work in the heart of the Marcellus Shale play. I am an engineer who has conducted numerous studies of the effects of UGE. I have studied the issue and have come to what I believe are important, necessary conditions that must be fulfilled before any locality decides to allow UGE in its neighborhood. Rather than spell these out, permit me to share a "what would it have been like if . . ." scenario. Here is a hypothetical letter exchange between the shale gas/oil industry and humans living over targeted shale formations. Please refrain from drawing conclusions until you read both letters.

To: Humans Worldwide Living over Targeted Shale Formations
From: The Shale Gas/Oil Industry

We are writing to ask your permission to develop shale gas/oil in your region using high-volume hydraulic fracturing from long horizontal well legs drilled from clustered, multiwell pads. Although you have allowed us to produce oil and gas from millions of wells over many years, we recognize that we are now asking you to allow us to do much more intense development than ever before, using a technology never before used in your area. We acknowledge our development plan for your region might eventually involve millions of new wells and be valued in the trillions of dollars, over decades to come.

We have seen how such intense development with this technology has caused problems where we are using it already. We have listened closely to your concerns about these problems, and others on the horizon, so we are writing you now to make a compact with you. We understand that you are granting us a privilege to give us the right to drill for gas/oil in your backyard. Quite honestly, our plans will significantly affect you, not just landowners or governments with whom we might have a business relationship. Therefore, if you give us the permission we seek, here are our promises to you:

1. We will not be developing in your area for another 2 to 3 years, so we have time to help you prepare for our arrival, as follows:
    - We will immediately fund appropriate training programs in your communities to produce home-grown workers for our industry. We will subsidize tuition for the students who commit to work in our industry. Those workers will get right of first refusal on our job openings.
    - We will immediately fund appropriate training programs for your emergency response teams—fire, police, medical, and spill hazards—and we will equip them at our expense.
    - We recognize that our heavy equipment will damage many of your roads and bridges. We will start now to pay to upgrade these so that they all remain usable not just by our equipment, but by you, too, throughout the development process. This will be a "stimulus" to help your unemployment situation now. When development is complete in an area, we will pay for final repairs necessary to leave all affected roads and bridges in state-of-the-art condition. This will be a legacy gift to you from our industry.
    - We will fund the construction or upgrading of regional industrial waste treatment and disposal facilities with adequate capacity to process safely all of the solid and liquid wastes we produce. We will not use sewage treatment plants to switch our garbage burden onto you. We will not truck your wastes to other regions for disposal.
2. We will be transparent about our entire plan for development:
    - We will tell you as soon as practicable, but no later than 1 year before start of activity, where and when we will drill, and what pipelines, compressor stations, and processing plants, and all other ancillary units will be needed where, and by when.
    - We will publish gas and waste production figures from every well, accurately, completely, and on time.
    - We will tell you where your oil/gas is going to market. We will not sell your oil/gas to a foreign market.
    - We will disclose, completely, all chemicals and other substances we use to you, not just to a regulatory agency.

3. We will accept, without debate, all new regulations that might be proposed by your regulatory agencies; your existing regulations are inadequate to cover the new technologies and cumulative impact of our industry. We will offer your agencies suggestions for continuous evolution of the regulations as a result of lessons we are learning.
4. With respect to your natural environment legacy:
   - For every tree we uproot, we will plant two replacements. We will reforest all access roads as quickly as we can and minimize the width of all forest cuts.
   - We will pay a fair price for the water we extract from your lakes and rivers, which will average many millions of gallons per gas well.
   - Whatever we break, despoil, or pollute, we will repair, replace, or remediate, at our expense.
5. We will safely dispose of all liquid and solid wastes from our development:
   - We will never store any flowback fluids or produced water in open pits. All such fluids will be recycled to the highest extent possible by existing technologies, regardless of increase in cost to us.
   - All liquid and solid wastes remaining from recycling will be treated at the aforementioned industrial waste treatment plants.
   - We will provide radiation monitoring equipment on every well pad: any materials, including drill cuttings, leaving a well pad that trigger an alarm will be sent to a licensed radioactive waste disposal facility.
6. We will not cause an increase in any tax levy on your citizens.
   - We will agree to a substantial increase in permit fees to reflect the expected fourfold increase in person-time we expect you to spend on review of permits for our industry.
   - We will agree to a state severance tax, the level of which will be floating, according to an accurate accounting of all costs to the state and municipalities.
7. We will practice what we preach about cleaner fuels and fewer emissions:
   - Every truck, every generator, every pump, every compressor will run on natural gas—no diesel or gasoline engines.
   - We will not allow gaseous emissions from any of our processes: no venting during flowback, no evaporation from open pits, no pressure releases from compressor stations or condensate tanks, no pipeline leaks.

8. We will be sensitive to noise and light pollution, even if a community does not have zoning restrictions in place to regulate these:
    - All of our pads and compressor stations will have sound/light suppression measures in place before start-up.
    - We will site drill pads, compressor stations, processing plants, pipelines, and all other ancillary units in collaboration with the community.
9. We will not unduly stress any of your communities:
    - We will never experiment with drilling many wells in a small area over a brief period of time. Dimock, Pennsylvania, should have taught us that lesson.
    - We will abide by all zoning restrictions on permitting.
    - We will never contest loss of well water use by any citizen. If a well is lost, we will replace it with whatever type of supply is requested by its owner at our expense, forever.
    - We will never require a citizen harmed by our development to promise silence in return for remediation.

Finally, and humbly, we note that even our best plans and efforts will come up short, sometime, someplace, somehow. *Therefore, in addition to all the contributions noted here, we also pledge to establish an escrow account that will receive 1% of the value of all gas/oil produced from our wells each year. This account will be administered by an independent third party, advised by an independent panel you select, and will be used as an emergency fund to compensate those financially or physically harmed by our development in your state.*

Thank you for your attention to our request.

How should individuals living over targeted shale formations respond? Of course each individual and each locality must come to their own conclusions about the benefits and the harms of UGE. Although many communities have embraced the industry and allowed drilling, others have fought, and are fighting, to block UGE in its backyard. A hypothetical response might read as follows:

To: The Shale Gas/Oil Industry
From: Humans Worldwide Living over Targeted Shale Formations

We have observed, calculated, thought, and done the science, and we have concluded that even "doing it right" is still wrong. We have a better plan for protecting our communities from the harms you admit, and our planet from proximate, dangerous climate change. We can't see the other end of your "bridge" fuel; rather, you ask us to walk a plank.

With the rights and privileges we possess to defend our communities, we politely answer: No thank you.

CONCLUDING THOUGHT

The issue is a hugely critical one that engenders strong emotions on both sides. The key question in my mind is this: Can the industry be trusted to get it right? After reading the chapters in this book, we hope you can come to an educated, informed decision. The stakes are very, very high—not just for the United States but also for the rest of the world. When it comes to UGE, we all do need to "get it right."

DISCLAIMER

My organization, Physicians, Scientists, and Engineers for Healthy Energy, Inc., provides an extensive online citation database of all aspects of shale gas and tight oil development at: http://psehealthyenergy.org/site/view/1180.

NOTE

1. *Colorado Judge Strikes Down Longmont's Fracking Ban in Favor of "State's Interest" in Oil and Gas.* http://ecowatch.com/2014/07/25/colorado-longmont-fracking-ban (accessed August 10, 2014).

About the Editor

MADELON L. FINKEL, PhD, is professor of health care policy and research and director of the Office of Global Health Education at the Weill Cornell Medical College in New York City. She is also course director in the Department of Healthcare Policy and Research. Her background is in epidemiology and health policy. Finkel's interests focus on women's health issues. Research projects include a cervical cancer screening and training program in rural Tamil Nadu, India. Over the past 4 years, she has focused on the health impacts of unconventional gas development on human health. She has written seminal pieces on the topic and has served as consultant to organizations involved with the issue. She has served as consultant to numerous law firms, pharmaceutical companies, and health care organizations on matters pertaining to epidemiology. She serves as secretary of the board of the Christian Medical College Vellore Foundation and is on the scientific advisory board for Physicians, Scientists, and Engineers for Healthy Energy, Inc. (PSE). Her book *Understanding the Mammography Controversy: Science, Politics and Breast Cancer Screening* was published by Praeger (2005). Praeger also published her book *Truth, Lies, and Public Health: How We Are Affected When Science and Politics Collide* (2007), and she served as editor of Praeger's three-volume text, *Public Health in the 21st Century* (2001). In 2014, Finkel became series editor for Praeger's series, Public Health Issues and Developments. This book is the first in that series on important and timely public health topics.

About the Contributors

JOHN L. ADGATE, PhD, MSPH, is professor and chair of the Department of Environmental and Occupational Health of the Colorado School of Public Health. His research interests include exposure assessment, risk assessment and environmental health policy, children's exposure to pesticides, asthma and allergen sampling, air pollution exposure analysis (volatile organic compounds and particulate matter), and exposure reduction interventions.

MICHELLE BAMBERGER, MS, DVM, is a veterinarian in private practice in Ithaca, New York, and serves on the advisory board of Physicians, Scientists, and Engineers for Healthy Energy. After receiving her degree from Cornell University's School of Veterinary Medicine, she studied at Oxford University and practiced small animal and exotic medicine and surgery in both Massachusetts and New York before opening Vet Behavior Consults in 2002. She has written several articles and a book on the subject of hydraulic fracturing and its impacts on human and animal health.

KATHRYN J. BRASIER, PhD, is an associate professor of rural sociology in the Department of Agricultural Economics, Sociology, and Education at The Pennsylvania State University. She previously was part of Penn State Extension's Economic and Community Development and Marcellus Education Teams. Her current research focuses on community impacts of unconventional energy development, network effects on learning and innovation in agriculture, and civic engagement in local environmental issues.

SUSAN CHRISTOPHERSON, PhD, is a professor and chair in the Department of City and Regional Planning at Cornell University and chair of the department. She is a geographer whose career has focused on urban and regional development and has published a series of key articles and a prize-winning book examining the influence of market governance regimes on regional economic development and firm strategies. She is also a recognized expert in the field of media studies illuminating the spatial dimensions of economy and society. Her recent work has been in the area of human–environment relations related to the development of unconventional sources of energy.

NATHAN P. DE JONG, MPH, is a research assistant in the Department of Environmental and Occupational Health of the Colorado School of Public Health. His research interests include exposure assessment and the health effects from oil and gas development.

MICHAEL H. DEPLEDGE, DSc, FSB, FRSA, is professor and chair of environment and human health, European Centre for Environment and Human Health, University of Exeter Medical School, Devon, United Kingdom. He has a background in environmental and medical toxicology and has been an expert advisor to the United Nations Environment programme and World Health Organization. He was formerly the chief scientist of the Environment Agency of England and Wales and served as a member of the Royal Commission on Environmental Pollution. He is also the former chairman of the Science Advisory Group on Environment and Climate Change of the Directorate General—Research and Innovation within the European Commission. He is currently a member of the UK government's Hazardous Substances Advisory Committee and the Global Health Committee of Public Health England. His research interests include the impact of climate change on health and well-being; the effects of environmental chemicals on human health, especially in the aging population; the use of the natural environment to promote health and well-being; and finding ways of communicating scientific information to policymakers and politicians.

DENNIS J. DEVLIN, PhD, is senior environmental health advisor, ExxonMobil Corporation, where he provides strategic guidance for environmental health policy and planning. He is a board trustee of the International Life Sciences Institute (ILSI) and past president of the ILSI Health and Environmental Sciences Institute, as well as chairman of the Petroleum Industry High Production Volume Testing Committee. He is currently chairman of the American Petroleum Institute's Exploration and Production Health Issues Group, which is an industry focal point for addressing the health issues related to unconventional resources development. He holds a PhD in toxicology.

MATTHEW FILTEAU, PhD, is a visiting assistant professor of sociology at Providence College, Providence, Rhode Island. His research examines the human dimensions of natural resource–based issues, including the social, economic, and environmental impacts from natural gas development in the Marcellus Shale region.

LIZ GREEN, MPH, is principal health impact assessment development officer at Public Health Wales and the Wales Health Impact Assessment (HIA) Support Unit in Wrexham, Wales, United Kingdom. Her work focuses on raising awareness of HIA, including developing, training, and advising on HIA in Wales.

JAKE HAYS, MA, is the director of the Environmental Health Program at Physicians, Scientists, and Engineers for Healthy Energy, Inc. (PSE) and a research associate at Weill Cornell Medical College, New York, New York. His principal focus has been on the environmental and public health aspects of unconventional oil and gas development. His research interests and formal training lie in the nexus of environmental law, policy, science, and ethics. He is currently pursuing a JD at Fordham University School of Law in New York City, where he serves as a board member of the Environmental Law Advocates and a staff member of the Environmental Law Review.

J. DAVID HUGHES is a geologist who has studied global and North American energy for four decades, including serving as research scientist and manager with the Geological Survey of Canada. He has published and lectured widely on long-term energy sustainability with a particular focus on unconventional fuels including shale. He is president of the consultancy Global Sustainability Research Inc. and a fellow of Post Carbon Institute. He also is a board member of the Physicians, Scientists, and Engineers for Health Energy, Inc. (PSE).

ANTHONY R. INGRAFFEA, PhD, PE, is the Dwight C. Baum Professor of Engineering and Weiss Presidential Teaching Fellow at Cornell University. His expertise is in rock mechanics, fracture mechanics, and computational science. He performed research and development and consulting for the oil and gas industry for more than 20 years. As a concerned scientist-engineer, he engages beyond the academy to further inform and educate the public on critical scientific issues that involve the public health and safety. He serves as treasurer of Physicians, Scientists, and Engineers for Healthy Energy, Inc. (PSE) in Ithaca, New York.

JONATHON KROIS, JD, is a project attorney with the Natural Resource Defense Council's (NRDC) New York Program. As part of NRDC's

Community Fracking Defense Project, he works with communities in New York, Pennsylvania, and across the nation to address the impacts of expanded natural gas development. He also works to advance recycling and producer responsibility, including electronic waste recycling, in New York. He worked with the Columbia Law School's Environmental Law Clinic on fracking, toxics regulation, and other environmental issues.

ADAM LAW, MD, is a clinical endocrinologist in Ithaca, New York, and founding member and president of Physicians, Scientists, and Engineers for Healthy Energy, Inc. (PSE) in Ithaca, New York. He is a fellow of the UK's Royal College of Physicians and is also a clinical assistant professor of medicine at Weill Cornell Medical College. He has written extensively on the health impacts of unconventional gas development, focusing on endocrine disrupting chemicals.

INMACULADA DE MELO-MARTÍN, PhD, MS, is professor of medical ethics in medicine at Weill Cornell Medical College. She holds a PhD in philosophy and an MS in molecular biology. Her research interests include bioethics and philosophy of science, and she has an extensive record of research, publication, and educational training in those areas. Her recent work focuses on the ethical issues of unconventional gas development.

ROBERT E. OSWALD, PhD, is professor of molecular medicine, College of Veterinary Medicine, at Cornell University. Research in his laboratory focuses on the structure and function of neurotransmitter receptors, in particular glutamate receptors. He is a prolific author, having published numerous articles and books on the impacts of unconventional gas development on human and animal health.

JEROME A. PAULSON, MD, is professor of pediatrics at the George Washington University School of Medicine & Health Sciences and professor of environmental & occupational health at the George Washington Milken Institute School of Public Health. He is the medical director for National & Global Affairs of the Child Health Advocacy Institute at the Children's National Medical Center where he also is the director of the Mid-Atlantic Center for Children's Health and the Environment, one of ten pediatric environmental health specialty units in the United States. He is the chairperson of the executive committee of the Council on Environmental Health American Academy of Pediatrics. He has also served on the Children's Health Protection Advisory Committee for the US Environmental Protection Agency.

DANIEL RAICHEL, JD, is a project attorney with the Natural Resources Defense Council's (NRDC) New York Program and a lead attorney with

NRDC's Community Fracking Defense Project, providing legal assistance to local governments, community groups, and individuals that are or may be harmed by the impacts of fracking or its associated infrastructure. Before joining NRDC, he worked on community and zoning issues related to gas drilling in the Columbia Law School Environmental Law Clinic.

SETH B.C. SHONKOFF, PhD, MPH, is executive director of Physicians, Scientists, and Engineers for Healthy Energy, Inc. (PSE) in Ithaca, New York, and visiting scholar in the Department of Environmental Science, Policy, and Management at the University of California at Berkeley. He has published extensively on topics related to air and water quality and the environmental and public health dimensions of energy and climate change from scientific and policy perspectives. Most recently, his work has focused on the human health dimensions of unconventional energy production, including shale gas and oil production in the United States and abroad.

KATE SINDING, JD, is a senior attorney and deputy director of the New York Program at the Natural Resources Defense Council (NRDC). She is also the director of NRDC's National Community Fracking Defense Project. In addition to her work to address the impacts of expanded oil and gas production nationwide, she works on advancing recycling programs involving the producer responsibility model, as well as other energy and land use matters. Before joining NRDC, she was a partner in the specialty environmental law firm of Sive, Paget & Riesel, PC. She has taught environmental law at Columbia University and Fordham University Schools of Law.

PHILIP L. STADDON, D.Phil, is an associate professor in environmental science at Xi'an Jiaotong-Liverpool University, Suzhou, China, and honorary senior research fellow in environment and health at the University of Exeter Medical School, Truro, United Kingdom. An ecologist by training, his background is in environmental change research focusing on the link between the environment and health and the role of ecology in determining health impacts. He is currently working on the impacts of climate change on human health and well-being.

VERONICA TINNEY, MPH, is the project coordinator for the Mid-Atlantic Center for Children's Health and the Environment at Children's National Health Center.

ROXANA Z. WITTER, MD, MSPH, MS, is an assistant research professor in the Department of Occupational and Environmental Health at the Colorado School of Public Health. She is the program director for the Occupational Medicine Residency Program. She is co-director of the Mountain and Plains Education and Research Center (MAP ERC) Occupational Health Interdisciplinary Symposium and has previously been director of the Occupational Toxicology MPH course. She is also a physician reviewer for the National Supplemental Screening Program for former Department of Energy Workers.

Index